ACCELERATOR PHYSICS - RADIATION SAFETY AND APPLICATIONS

Edited by **Ishaq Ahmad** and **Maaza Malek**

Accelerator Physics - Radiation Safety and Applications
http://dx.doi.org/10.5772/intechopen.68264
Edited by Ishaq Ahmad and Maaza Malek

Contributors

Hai Lin, Oyeon Kum, Vladimir E. Teryaev, Joseph Orgel, Rama Madhurapantula, Serge Y. Kalmykov, Xavier Davoine, Isaac Ghebregziabher, Bradley Shadwick, Alexander Barzilov, Matthew Hodges, Karen Cloete, Nikolay Vasilievich Smolyakov

Published by InTech
Janeza Trdine 9, 51000 Rijeka, Croatia

Publishing Process Manager Iva Simcic
Technical Editor SPi Global
Cover InTech Design team

First published February, 2018
Printed in Croatia
Legal deposit, Croatia: National and University Library in Zagreb

Additional hard copies can be obtained from orders@intechopen.com

Accelerator Physics - Radiation Safety and Applications, Edited by Ishaq Ahmad and Maaza Malek
p. cm.
Print ISBN 978-953-51-3835-8
Online ISBN 978-953-51-3836-5

For EU product safety concerns:
IN TECH d.o.o., Prolaz Marije Krucifikse Kozulić 3, 51000 Rijeka, Croatia,
info@intechopen.com or visit our website at intechopen.com.

Meet the editors

Ishaq Ahmad joined the National Centre for Physics in 1999 and currently holds the position of Director, Experimental Physics Department, National Centre for Physics, Islamabad, and is an adjunct faculty member of Pakistan Institute of Engineering and Applied Sciences (*PIEAS*), Islamabad. He has coauthored over 80 publications in SCI journals and supervised several master, PhD, and postdoctoral students. Dr. Ahmad has authored a number of books and book chapters. His research interests are focused on ion implantation, radiation-induced modification of materials/nanomaterials, synthesis of nanomaterials/thin films, and ion beam analysis of materials. He is an editorial board member and reviewer of several well-known journals. Dr. Ahmad was a TWAS-UNESCO-iThemba LABS research associate from 2014 to 2016. He obtained his PhD degree at Graduate University of Chinese Academy of Sciences, China.

Prof. M. Maaza holds a PhD degree in Quantum Neutron Optics from the University of Pierre-Marie Curie and the Commissariat a l'Energie Atomique, respectively. He is currently a permanent joint staff member of the National Research Foundation of South Africa, in charge of the Africa-International relations desk at iThemba LABS and the University of South Africa (UNISA). Prof. M. Maaza is the chair of reputed international organizations and commissions. His expertise is in the multidisciplinary field of nanosciences, photonics, and solar energy. Within the STI and RDI landscapes, with a Hirsh index of 33, Prof. M. Maaza has published more than 300 ISI international publications and supervised several postgraduates from all over Africa and the South.

Contents

Preface

Today accelerator physics is one of the most promising areas of science and technology to observe the secret of matter as well as the origin of nature which create new opportunities for accelerator physics research. This book provides a glance view at recent research on accelerator physics, technologies, as well as the use of accelerators in research and biology. The main aim of this book is to address latest issues in accelerator physics and to review the most recent applications of ion beam as well as synchrotron radiation.

The first part of this book covers the relevant issues in beam dynamics. A universal method on dynamics of electron beam described by well-known Vlasov-Maxwell (V-M) equations solves the bottleneck of numerical study on 6D phase space dynamics and leads to exact solutions of the 3D macroscopic dynamics of a beam through available computer group condition. Motion of relativistic electrons in three-dimensional ideal undulator magnetic field is also presented in this book.

The review articles in the second part cover the recent achievements on advanced accelerator design. The concept and numerical simulation of high efficiency electron multi-beam linear accelerator EVT for industrial and medical applications is reviewed in this book. Recent research on optically controlled laser-plasma electron acceleration for compact gamma-ray sources is also reviewed.

Safety assessment is a systematic and comprehensive methodology to evaluate risks associated with a complex technological entity such as a high-energy heavy-ion medical accelerator facility. The third part of this book covers the nuclear safety aspects such as radiation shielding and activation analysis for the preliminary design of heavy-ion medical accelerator facility as well as radiation safety aspects of operation of electron linear accelerators equipped with bremsstrahlung converters.

The last part is devoted to the most important applications of accelerators. Today synchrotron radiation is an enormously *versatile tool for* various kinds of spectroscopy, structure analysis, and a number of other applications in different fields of sciences and technology. The determination of the structure of biomolecules is presently among the best examples of the application of synchrotron radiation. This book also covers synchrotron-based X-ray diffraction study of mammalian connective tissues and related disease. Furthermore, an overview of the versatile applications of ion beam and synchrotron radiation techniques in hair elemental profiling in biomedical studies is incorporated.

Finally and foremost, we wish to record our sincere appreciation to the researchers who potentially contributed to this book: Prof. Dr. H. Lin, Prof. Dr. Nikolay Smolyakov, Prof. Dr. Vladimir E. Teryaev, Prof. Dr. Serge Y. Kalmykov, Prof. Dr. Xavier Davoine, Prof. Dr. Isaac

Ghebregziabher, Prof. Dr. Bradley A. Shadwick, Prof. Dr. Oyeon Kum, Prof. Dr. Matthew Hodges, Prof. Dr. Alexander Barzilov, Prof. Dr. Rama Sashank Madhurapantula, Prof. Dr. Joseph P.R.O. Orgel, and Ms. Iva Simcic, Publishing Process Manager.

Dr. Ishaq Ahmad
National Centre for Physics
Pakistan

Prof. Maaza Malek
UNESCO-UNISA Africa Chair in Nanosciences/Nanotechnology
UNISA and Nanosciences African Network
iThemba LABS-National Research Foundation
South Africa

Section 1

Beam Dynamics

Chapter 1

Phase Space Dynamics of Relativistic Particles

Hai Lin

Additional information is available at the end of the chapter

http://dx.doi.org/10.5772/intechopen.70957

Abstract

By analyzing bottleneck of numerical study on six-dimensional (6D) phase space dynamics of electron beam, we present a universal and practical scheme of exactly simulating the 3D dynamics through available computer group condition. In this scheme, the exact 6D phase space dynamics is warranted by exact solutions of 3D self-consistent fields of electron beam.

Keywords: Vlasov-Maxwell system, electron beam, phase space dynamics, microscopic distribution function, self-consistent fields

1. Introduction

The dynamics of electron beam in various man-made apparatus is a kernel content of accelerator/beam physics [1–11]. Such a beam can be described by well-known Vlasov-Maxwell (V-M) equations [1–3, 12–14]. Because usually the number of realistic particles (each has a mass m_e and a charge e) in an electron beam is at astronomical figure level, it is really a huge challenge to calculate realistic 3D dynamics of such a beam even through the most advanced computer groups available currently.

This can be well illustrated by following rough estimation. An electron needs to be described by six real-valued variables (for its three position components and three velocity components), and each real-valued variable will cost ~24 byte storage mount. For a realistic beam with a total charge at pC level, the number of realistic particles it contains will be about $10^{-12+19}/1.6$~6.3×10^7 and hence corresponds to 1.5 GB storage mount. If numerical experiment is conducted through a single-CPU computer, according to the most advanced personal computer currently available, 1.5 GB can be contained in the memory which is commercially available at 4 GB level. Thus, if all data can be stored in the memory, time-consuming data exchange between the memory, a high-speed storage medium, and the disk, a low-speed storage medium, can be avoided and the time cost of updating these data per round will be determined by the time cycle of CPU, which is at *ns* level. Because updating a real-valued variable will spend several cycles (and hence tens *ns*), the time cost for updating 1.5 GB data recording 6.3×10^7 real-valued variables will be at minute level (if not taking into account the time cost for updating fields information). Because the time step is chosen to be sufficiently small even for explicit difference scheme, usually at least updating hundreds of times is needed for obtaining a snapshot of the beam at a specified time point. This implies at least hour-level time cost through the above-described most advanced computer condition currently. Of course, the larger the total charge is, the more the time cost is (if the data for the beam is still able to be contained in a memory).

Above discussion has clearly revealed that only for low-charge beam, the numerical experiment, which is based on explicit difference scheme, can be conducted on a single-CPU computer. Therefore, for more general cases with high-charge beam, people have to resort to computer group and parallel code. In such a situation, if data are not exchanged between high-speed memory and low-speed disk, the bottleneck of the experiment is not the time cycle of a CPU but that of the main board which might be several folds of that of the CPU and affect the dispatch of data bus responsible for data dispatch among CPUs. Sufficient CPUs can offset the effect of the time cycle of the main board. But the fact that the larger the total charge is, the more the time cost is still exists. Thus, for *nC*-level beam and stricter implicit-difference-scheme of updating fields, CPUs' number in thousand levels is still unable to do as one would wish.

The contradiction between feasible computers' condition and large amount of data describing realistic beams often force researchers to make some compromises. Merging $N>1$ realistic particles (each is of a mass m_e and a charge e) into a macroparticle (of a mass Nm_e and a charge Ne) is a common compromise. Merging can cause less data mount for describing a beam and hence can speed up numerical experiment. Therefore, it is broadly adopted in popular scheme of simulating particle dynamics, such as particle-in-cell (PIC) simulation [15, 16].

Although merging can speed up the numerical experiment, it is at the cost of the reliability of the experiment. Let us analyze what will happen when $N=2$. If there are l realistic particles in a beam, describing the beam will need $6l$ functions: $\{x_i(t), y_i(t), z_i(t), d_t x_i(t), d_t y_i(t), d_t z_i(t)\}$ and $1=<i=<l$. Usually, the merging is based on the following principle: at $t=0$, each realistic particle merges the nearest neighbor particle. Thus, no matter how large the difference between the initial velocities of any two merged particles, denoted as $d_t x, y, z_{2k}(t=0) - d_t x, y, z_{2k+1}(t=0)$, is, $d_t x, y, z_{2k} - d_t x, y, z_{2k+1}$ is always taken as 0 or the $2k$th realistic particle is always bound with the $2k+1$th one (their relative motion is ignored completely). This is a distorted description. The $2k$th realistic particle and the $2k+1$th one, if they have the same initial positions $x, y, z_{2k}(t=0) = x, y, z_{2k+1}(t=0)$, tend to move to different destinations at the next time point $0+\Delta t$ (where Δt is the time step), because of their different initial velocities $d_t x, y, z_{2k}(t=0)$ and $d_t x, y, z_{2k+1}(t=0)$. Namely, actual picture should be described as follows: at every time point t, a macroparticle will break into N equal pieces having their own destinations at $t+\Delta t$, rather than globally move into a destination at $t+\Delta t$ or being equivalently expressed as: each macroparticle at t is a gathering of N equal pieces from different sources at $t-\Delta t$. Because of

the distorted description, we briefly summarize the merging as the rigid-macroparticle approximation which refers to a macroparticle having a fixed mass Nm_e and a fixed charge Ne.

The direct result of the rigid-macroparticle approximation is that the charge density profile n of the beam at next time point $t+\Delta t$ is distorted, and hence the self-consistent electric field E (meeting $\nabla \cdot E = ne$) is inexact. If the dynamics calculation cannot be satisfactory even over a short time scale down to the time step Δt, it is hopeless to expect calculation over longer time scale which is multifold of Δt to be reliable.

Actually, even if computer condition is sufficiently advanced to directly calculate realistic particles, Newton equations (NEs) and Maxwell equations (MEs), complicated mathematical relation between E and B, which is represented by

$$\partial_t E + \nabla \times B = \mu_0 j, \tag{1}$$

leads to considerable difficulty in setting up difference scheme of each NE

$$d_{tt} r_i(t) = \frac{e}{m_e} [E(r_i(t), t) + d_t r_i(t) \times B(r_i(t), t)]. \tag{2}$$

This is because that at any time t, although E can be known from all particles' information at t, B cannot. Due to the term $\partial_t E$ in Eq. (1), the information of B at t have dependence on all particles' information at $t+\Delta t$. This implies that the NE is a complicated differential equation of $r_i(t)$, and its difference scheme is difficult to be established. Namely, each difference version of Newton equation involves indeed all particles' coordinates at $t+\Delta t$, i.e., $\{r_i(t+\Delta t); 1 =< i =< N\}$. Thus, N difference versions form a linear equation set of $\{r_i(t+\Delta t); 1 =< i =< N\}$ whose solutions involve inevitably a $N \times N$ matrix. The larger N is, the more hopeless exact solution is.

Clearly, an exact or reliable dynamics calculation on any V-M system needs more in-depth exploring its universal properties, which can be favorable to avoid above-mentioned contradictions. In the following paragraph, we reveal some universal properties of the V-M system. Based on these universal properties, we establish a universal and practical scheme of exact dynamics calculation of any V-M system.

2. Universal properties of Vlasov-Maxwell system

It is well known that Vlasov equation (VE) $\widehat{L}f = 0$, where the operator is $\widehat{L} = [\partial_t + v \cdot \nabla - LF_v \cdot \partial_p]$ and $LF_v = e[E(r,t) + v \times B(r,t)]$ represents the Lorentz force and $p(v) = v\Gamma(v)$ and $\Gamma(v) = \frac{1}{\sqrt{1 - v \cdot v}}$ correspond to the conservation law of total particle number $\int f d^3 r d^3 v = cons\tan t$. We should notice an important fact: if a beam is a particle number conserved system, its any portion or subsystem is not always also particle number conserved. For an unconserved subsystem, its distribution function g does not meet the VE. An unconserved subsystem means there are always realistic particles leaving it. A macroparticle is analogous to such an unconserved subsystem because it will break into equal pieces having different destinations at next time point $t+\Delta t$. Namely, there are always pieces leaving the macroparticle.

According to Maxwell equations (MEs), (E, B) couples with (M_0, M_1), where $M_i = \int v^i f d^3 v$, and formally decouples with all $M_{i>=2}$. According to open fluid equations set $\left\{\int v^{i>=0} \widehat{L} f d^3 v = 0, 0 =< i < \infty\right\}$, each equation $\int v^i \widehat{L} f d^3 v = 0$ reflects a universal relation among $\partial_t M_i$, ∇M_{i+1} and at least M_{i-1}. This seems that exact solutions of all higher order moments $M_{i>=2}$ are necessary for that of (E, B). On the other hand, because $p(v)$ and $\Gamma(v)$ are nonlinear functions of v and hence $-\int v^i L F_v \cdot \partial_p f d^3 v$, a term in the equation $\int v^{i>=1} \widehat{L} f d^3 v = 0$, is dependent on moments M_m from $m = i$ to $m = \infty$, each equation $\int v^{i>=1} \widehat{L} f d^3 v = 0$ will imply a relation among moments M_m from $m = i$ to $m = \infty$, and thus all equations $\int v^i \widehat{L} f d^3 v = 0$ from $i = 1$ to $i = \infty$ will mean a $\infty \times \infty$ matrix describing the relation among all moments M_m from $m = 1$ to $m = \infty$. Clearly, obtaining exact solutions of (E, B) from such an open equation set $\left\{\int v^{i>=0} \widehat{L} f d^3 v = 0, 0 = < i < \infty\right\}$ is impossible and, as discussed below, also unnecessary.

Another open set $\{D_i, 0 = < i < \infty\}$, where $D_i = \left[\frac{M_i}{M_0} - \left(\frac{M_1}{M_0}\right)^i\right]^* M_0$, can be defined naturally through the M-set. Clearly, $D_0 = 0$ and $D_1 = 0$ automatically exist. Each equation $\int v^i \widehat{L} f d^3 v = 0$ can be expressed through the D-set

$$A_i \partial_t D_i + B_{i+1} \nabla D_{i+1} + \sum_{m>=i} C_m D_m = 1, \tag{3}$$

and coefficients A_i, B_i, C_i are known functionals of (E, B, M_0, M_1). Starting from the $i = 1$ case, we can formally obtain an expression of D_2 in all terms $D_{i>=3}$ and then substitute it into the $i = 2$ case and formally obtain an expression of D_3 in all terms $D_{i>=4}$, etc. Finally, we will find that all $D_{i>=2}$ are determined by D_∞ and all coefficients A_i, B_i, C_i. Namely, the open equation set $\left\{\int v^i \widehat{L} f d^3 v = 0, 1 =< i < \infty\right\}$ does not lead to a substantial constraint on (E, B, M_0, M_1).

According to MEs, (E, B) depends on (M_0, M_1) and is independent of the D-set. Therefore, the exact solutions of the D-set are not a necessary condition for those of (E, B). The open equation set $\left\{\int v^i \widehat{L} f d^3 v = 0, 1 =< i < \infty\right\}$ is only responsible for relations among those D_i and cannot have an effect on (E, B, M_0, M_1).

Two universal properties of any V-M system are favorable to achieve exact dynamics calculation. One is that the f of any V-M always has a thermal spread [18], or in stricter mathematical language, functions in a general form $g(r, t) * \delta(v - u(r, t))$, where g and u are two functions of (r, t) and δ is Dirac function, cannot meet the VE or $\widehat{L}[g(r, t) * \delta(v - u(r, t))] \neq 0$. The other is that the open set $\{M_i, 0 = < i < \infty\}$ has a closed subset $\{M_0, M_1\}$, which meets a closed equation set. Detailed proof of the second universal property is presented below.

Due to the first universal property, the whole system can be viewed as a superposition of a cold subsystem denoted by $F(f) = \delta\left(v - \frac{\int f v d^3 v}{\int f d^3 v}\right) * \int \left[f * \delta\left(v - \frac{\int f v d^3 v}{\int f d^3 v}\right)\right] d^3 v$, where δ is Dirac function, and a thermal subsystem denoted by $f - F(f)$. The cold subsystem inherits a part of each moment $\int v^i f(r, v, t) d^3v$, so does the thermal subsystem. Namely, the cold subsystem has its own self-consistent field (E^F, B^F), where $\nabla \cdot E^F \sim \int F(r, v, t) d^3v$ and $\nabla \times B^F \sim \int vF(r, v, t) d^3v$. Likewise, the thermal subsystem has its own self-consistent field (E^{f-F}, B^{f-F}), where $E = E^F + E^{f-F}$ and $B = B^F + B^{f-F}$. (E^F, B^F) plays a role of external fields to the thermal subsystem and so does (E^{f-F}, B^{f-F}) to the cold subsystem. In particular, two subsystems have a common fluid velocity: $u = \frac{\int vF(r,v,t)d^3v}{\int F(r,v,t)d^3v} = \frac{\int v[f-F](r,v,t)d^3v}{\int v[f-F](r,v,t)d^3v} = \frac{\int vf(r,v,t)d^3v}{\int vf(r,v,t)d^3v}$. Because F has a known dependence on v, its higher order moments $\int v^{i>1} F d^3v$ are of simple forms $u^{i>1} * \int [f*\delta(v-u)] d^3v$. The thermal subsystem inherits the thermal spread of f and hence all off-center moments, the dependence of $f-F$ on v is still unknown, and its higher-order moments $\int v^{i>1}[f-F] d^3v$ still obey a set of equations in infinite number. The equations of these higher-order moments $\int v^{i>1} F d^3v$ in infinite number are found, as proven later, to be equivalent to an equation of $u = \frac{\int vF(r,v,t)d^3v}{\int F(r,v,t)d^3v}$.

Detailed strict proof has been made elsewhere [17, 18] (see the appendix of Ref. [17]).

Here, for convenience of readers, we briefly present the kernel of strict proof. For simplicity of symbols, we denote $\left[\widehat{L} + (v \cdot \nabla u)\partial_v\right]F$ as Ω and F as $n_0\delta(v-u)$, where $n_0 \equiv \int [f*\delta(v-u)] d^3v < \int f d^3v = M_0$. From the definition of F, there is always $u = \frac{\int vFd^3v}{\int Fd^3v}$. Because of strict mathematical formulas $\int d_v\delta(v-u)d^3v = 0$, $\int [v-u]*\delta(v-u)d^3v = 0$, $\int [v-u]*[v \cdot \nabla + (v \cdot \nabla u)\partial_v]\delta(v-u) d^3v = 0$, and $\int [v-u]*W(r, v, t)*d_v\delta(v-u)d^3v = \int [v-u]*W(r, v, t)|_{v=u}*d_v\delta(v-u)d^3v$, where $W(r, v, t)$ is the arbitrary function, the definition of Ω can be directly written as

$$\Omega = [\partial_t + u \cdot \nabla] n_0 * \delta(v-u) - n_0 * \left[\partial_t u + (LF_v * \partial_p v)\big|_{v=u}\right] * \delta' = \left\{\int \Omega d^3v\right\} * \delta - \left\{\int [v-u] * \Omega d^3v\right\} * \delta', \quad (4)$$

which has a strict solution $\Omega = 0$. Therefore, there is a theorem: For any V-M system, its $F(f)$ meets $\Omega = 0$. Namely, $\widehat{L}f = 0$ means $\Omega = 0$.

Due to $0 = \int v^0 \Omega d^3v$, $\int v^1 \Omega d^3v = 0$ will be be equivalent to [17, 18]

$$\partial_t[p(u)] + e[E + u \times B] = 0. \quad (5)$$

For equations in infinite number $0 = \int v^{i>=2} \Omega d^3v$, the Dirac function dependence of F on v makes all these equations indeed to be equivalent to a same equation, Eq. (5). This can be easily verified by simple algebra.

Few authors use Eq. (5) to calculate (E, B) in plasmas [17–22]. As previously pointed out, the equation $\int v\widehat{L}fd^3v = 0$ reflects a relation among all $D_{i>=2}$. To determine these $D_{i>=2}$, the whole set $\left\{\int v^{i>=0}\widehat{L}fd^3v = 0, 0 =< i < \infty\right\}$ is required.

The whole system is described by

$$\widehat{L}f = 0; \tag{6a}$$

$$\partial_t E = -[\nabla \cdot E - ZeN_i]\frac{\int vfd^3v}{\int fd^3v} + \nabla \times B; \nabla \cdot B = 0 \tag{6b}$$

$$\nabla \times E = -\partial_t B. \tag{6c}$$

Likewise, the cold subsystem by

$$\left[\widehat{L} + (v \cdot \nabla u)\partial_v\right]F = 0; \tag{7a}$$

$$\partial_t E = -[\nabla \cdot E - ZeN_i]\frac{\int vFd^3v}{\int Fd^3v} + \nabla \times B; \nabla \cdot B = 0 \tag{7b}$$

$$\nabla \times E = -\partial_t B, \tag{7c}$$

and the thermal subsystem by

$$\left[\widehat{L}(f - F) - (v \cdot \nabla u)\partial_v F\right] = 0; \tag{8a}$$

$$\partial_t E = -[\nabla \cdot E - ZeN_i]\frac{\int v(f - F)d^3v}{\int (f - F)d^3v} + \nabla \times B; \nabla \cdot B = 0 \tag{8b}$$

$$\nabla \times E = -\partial_t B. \tag{8c}$$

Two universal properties imply a universal scheme of exact dynamics calculation of any V-M system. That is, the VE $\widehat{L}f = 0$ implies $\Omega = 0$, which implies $\{\int v^i \Omega d^3v = 0, 0 = < i < \infty\}$, which implies a relation between $\frac{\int v^1 Fd^3v}{\int v^0 Fd^3v}$ and (E, B), which is also a relation between $\frac{\int v^1 fd^3v}{\int v^0 fd^3v}$ and (E, B). On the other hand, with the help of the CE $\int v^0 \widehat{L}fd^3v = 0$, the open equation subset $\left\{\int v^i \widehat{L}fd^3v = 0, 1 = < i < \infty\right\}$ will be equivalent to a $\infty \times \infty$ matrix equation which gives exact expressions of all higher order off-center moments $D_{i>=2}$ in terms of (E, B, M_0, M_1). After obtaining exact solutions of (E, B, M_0, M_1) from Eq. (2) and MEs, all $D_{i>=2}$ can be known through $\left\{\int v^i \widehat{L}fd^3v = 0, 1 = < i < \infty\right\}$.

This universal scheme gets rid of inevitable shortcomings of other methods. For example, well-known PIC simulation is to introduce macroparticle representation of initial distribution function and to calculate coupled evolution of these macroparticles and E and B. As analyzed above, such a macroparticle is unrealistic because it approximates that all contained realistic particles always stay in it and completely ignore the fact that realistic particles can leave it. Moreover, because the coupled evolution is calculated in an alternatively updating manner, this makes calculated E and B to be dependent of the grainness parameter which reflects how many realistic particles are contained in a macroparticle and hence are not exact solutions of E and B. Thus, even if f can be expressed through those macroparticles' positions and velocities, inexact solutions of E and B make such an expression of f to be inexact.

Therefore, even though advanced computer group is applied to PIC simulation, its abovementioned inherent shortcoming cannot warrant more advanced computer condition corresponding to Exacter solutions of f (because E and B are inexact solutions). Namely, advanced computer condition does not get reasonable usage. In contrast, if the phase space dynamics is conducted according to above-described manner based on six-equation set, the power of the advanced computer condition can be fully utilized to get higher resolution of exact solutions of f (because E and B are exact solutions).

3. Universal properties of Newton-Maxwell system

Eq. (5) can also be derived from Newton-Maxwell (N-M) system, which contains N relativistic Newton equations (RNEs) and four MEs. Actually, for a group of electrons $\{r_i, d_t r_i\}$, we can always define two fields (in Lagrangian expression) [19]:

$$u(r_i(t),t) \equiv \frac{\sum_j d_t r_j \delta\big(r_i(t) - r_j(t)\big)}{\sum_j \delta\big(r_i(t) - r_j(t)\big)}, \tag{9}$$

$$RV(r_i(t),t) \equiv \frac{\sum_j \big[d_t r_j - d_t r_i\big] \delta\big(r_i(t) - r_j(t)\big)}{\sum_j \delta\big(r_i(t) - r_j(t)\big)}, \tag{10}$$

which are fluid velocity field and relative velocity field, respectively. Especially, there is always a formula

$$d_t r_i \equiv u(r_i(t),t) - RV(r_i(t),t), \tag{11}$$

where $\equiv$ means the formula is automatically valid for any $r_i(t)$. If applying $\nabla_{r_i(t)}$ to this automatically valid formula, we will obtain another automatically valid formula:

$$\nabla_{r_i(t)}[u(r_i(t),t) - RV(r_i(t),t)] \equiv \nabla_{r_i(t)} d_t r_i \equiv d_t \nabla_{r_i(t)} r_i \equiv d_t 1 \equiv 0. \tag{12}$$

Because N electrons are described by four MEs and N RNEs, this automatically valid formula enables each RNE to be written as

$$\begin{aligned} 0 &= d_t p(d_t r_i) + eE(r_i(t),t) + e d_t r_i \times B(r_i(t),t) = d_t p(u(r_i(t),t) - RV(r_i(t),t)) + eE(r_i(t),t) \\ &+ e[u(r_i(t),t) - RV(r_i(t),t)] \times B(r_i(t),t) = \partial_t p(u(r_i(t),t) - RV(r_i(t),t)) + d_t r_i \times \nabla_{r_i(t)} p[u(r_i(t),t) \\ &- RV(r_i(t),t)] + eE(r_i(t),t) + e[u(r_i(t),t) - RV(r_i(t),t)] \times B(r_i(t),t) = \partial_t p(u(r_i(t),t) - RV(r_i(t),t)) \\ &+ eE(r_i(t),t) + e[u(r_i(t),t) - RV(r_i(t),t)] \times B(r_i(t),t), \end{aligned} \tag{13}$$

where $p(\text{var}) = \frac{\text{var}}{\sqrt{1-var*\text{var}}}$ is relativistic momentum, because no matter what the relative field is, the RNE of every electron is always valid. Thus, the condition for the RNE being valid under arbitrary value of the *RV* field, of course including a common value $RV=0$, is that the following equation is valid:

$$0 = \partial_t p(u(r_i(t), t)) + eE(r_i(t), t) + e[u(r_i(t), t)] \times B(r_i(t), t). \quad (14)$$

Its Euler expression reads

$$0 = \partial_t p(u(r, t)) + eE(r, t) + e[u(r, t)] \times B(r, t), \quad (15)$$

which is just Eq. (5). This motion equation of $u(r, t)$ and four MEs form a closed five-equation set. No matter how large N is, it merely means more *RV*-field fluid elements being calculated. Finer description of the *RV* field will lead to Exacter information on thermal distribution, but the *RV*-field has no contribution to the electric current and hence E and B. Thus, the exact solutions of the $N+4$ equation set (N RNEs and four MEs) can be obtained from an $N+5$ equation set (N equations of $RV(r_i(t), t)$ and a closed five-equation set of (u, E, B). Examples of electron density profile based on this closed five-equation set have been presented elsewhere [17–20].

4. Examples of exact solution of phase space distribution

From Eq. (5) and MEs, we can find that $W \equiv \partial_t p(u) + ZN_i r/3$ and $u \times B$ meeting a same equation [17, 18]. Further strict analysis reveals that this will lead to a self-consistent equation of u [17, 18]:

$$\partial_{tt} p + \left[(\partial_t \nabla \cdot p) \frac{p}{\sqrt{1+p^2}} \right] + \nabla \times \nabla \times p = -[uZN_i] + POT \quad (16)$$

where Z is ionic charge, N_i is ionic density, and *POT* is a constant vector determined by initial conditions. After solving $u(r, t)$ from this equation, $B(r, t)$ will be known from a formula $\partial_t p + ZN_i r/3 = -\lambda[u \times B]$ [17, 18], where $\lambda \neq 1$ is a constant, and then $E(r, t)$ from $E = \partial_t p + u \times B$ [17, 18].

Here, we give some typical exact solutions of the distribution of f over the v-space [17, 19–21]. If f is an exact solution of the VE, any monovariable function of f, or $g(f)$, is also an exact solution. Because Eq. (5) and MEs can have exact solutions of (E, B) in running-wave form, $E = E\left(r - \frac{1}{\eta} ct\right)$ and $B = B\left(r - \frac{1}{\eta} ct\right)$, we should note a relation between E and B, $E = -\frac{1}{\eta} c \times B + \nabla\Phi\left(r - \frac{1}{\eta} ct\right)$, which arises from $\nabla \times E = -\partial_t B$. Here, $\Phi\left(r - \frac{1}{\eta} ct\right)$ is a scalar function but cannot be simply taken as electrostatic (ES) potential (because $-\frac{1}{\eta} c \times B$ also has divergence or $\nabla \cdot \left(-\frac{1}{\eta} c \times B\right) = \frac{1}{\eta} c \cdot \nabla \times B \neq 0$). In this case, the VE can be further written as

$$0 = \left[\partial_t + v\cdot\nabla - [E + v\times B]\cdot\partial_{p(v)}\right]f = \left[\partial_t - E\cdot\partial_{p(v)}\right]f + v\cdot\left[\nabla - B\times\partial_{p(v)}\right]$$

$$f = \left[\partial_t + \frac{1}{\eta}c\times B\cdot\partial_{p(v)}\right]f + v\cdot\left[\nabla - B\times\partial_{p(v)}\right]f - \nabla\Phi\cdot\partial_{p(v)}$$

$$f = \left[\partial_t + \frac{1}{\eta}c\cdot B\times\partial_{p(v)}\right]f + v\cdot\left[\nabla - B\times\partial_{p(v)}\right]$$

$$f - \nabla\Phi\cdot\partial_{p(v)}f = \left(v - \frac{1}{\eta}c\right)\cdot\left[\nabla - B\times\partial_{p(v)}\right]f - \nabla\Phi\cdot\partial_{p(v)}f. \tag{17}$$

It is easy to verify that any function of $p + \int E\left(r - \frac{1}{\eta}ct\right)dt$ will meet

$$0 = \left[\partial_t - E\cdot\partial_{p(v)}\right]g\left(p + \int E\left(r - \frac{1}{\eta}ct\right)dt\right) = \left[-\frac{1}{\eta}c\cdot\nabla + \frac{1}{\eta}c\cdot B\times\partial_{p(v)}\right]g = -\frac{1}{\eta}c\cdot\left[\nabla - B\times\partial_{p(v)}\right]g, \tag{18}$$

where we have used the property $\partial_t\int E\left(r - \frac{1}{\eta}ct\right)dt = -\frac{1}{\eta}c\cdot\ \nabla\int E\left(r - \frac{1}{\eta}ct\right)dt$. Thus, if $\nabla\Phi\equiv 0$, any mono-variable function of $p + \int E\left(r - \frac{1}{\eta}ct\right)dt$, or $g\left(p + \int E\left(r - \frac{1}{\eta}ct\right)dt\right)$, will be an exact solution of the VE.

On the other hand, for more general $\nabla\Phi$, we can find that any monovariable function of $\sqrt{1 + \frac{p^2}{c^2}} - \frac{1}{\eta}c\cdot p + \Phi$ or $g\left(\sqrt{1 + \frac{p^2}{c^2}} - \frac{1}{\eta}c\cdot p + \Phi\right)$, is an exact solution of the VE. According to Eq. (14), $\partial_p\left[\sqrt{1 + \frac{p^2}{c^2}} - \frac{1}{\eta}c\cdot p\right]$ will contribute a vector parallel to $\left(v - \frac{1}{\eta}c\right)$ and hence make the operator $\left(v - \frac{1}{\eta}c\right)\cdot B\times\partial_{p(v)}$ has zero contribution.

Therefore, for running-wave form, $E = E\left(r - \frac{1}{\eta}ct\right)$ and $B = B\left(r - \frac{1}{\eta}ct\right)$, the phase space distribution, if $\nabla\Phi\equiv 0$, can be described by a positive-valued function of $p + \int E\left(r - \frac{1}{\eta}ct\right)dt$, for example, $\exp\left[-\left(p + \int E\left(r - \frac{1}{\eta}ct\right)dt\right)^2\right]$, $\sin^2\left(\exp\left[-\left(p + \int E\left(r - \frac{1}{\eta}ct\right)dt\right)^2\right]\right)$, etc. We can further pick out reasonable solutions according to the running-wave form of u:

$$u = \frac{\int \frac{p}{\sqrt{1+p^2}} g\left(p + \int E\left(r - \frac{1}{\eta}ct\right)dt\right)d^3p}{\int g\left(p + \int E\left(r - \frac{1}{\eta}ct\right)dt\right)d^3p}. \tag{19}$$

Likewise, same procedure exists for more general $\nabla\Phi$ and $g\left(\sqrt{1 + \frac{p^2}{c^2}} - \frac{1}{\eta}c\cdot p + \Phi\right)$.

Actually, a set of macroscopic functions (E, B, u) can have multiple microscopic solutions of the VE. Therefore, usually we know the distribution of f over the v-space from the initial condition

$f(r,p,t=0)$. From the function dependence of $f(r,p,t=0)$ on p, we can obtain the function dependence of g on $\sqrt{1+\frac{p^2}{c^2}}-\frac{1}{\eta}c\cdot p$ and hence determine detailed function form of g.

Detailed procedure of determining the function form of g is described as follows [9, 11]: we can seek for special space position R in which $E(R,0)=-\frac{1}{\eta}c\times B(R,0)$, or $\nabla\Phi(r,0)|_{r=R}=0$, exists. The initial distribution at R, i.e., $f(R,p,t=0)$, is thus a monovariable function p. At the same time, two expressions are equivalent, and hence there is $f(R,p,t=0)=g(K+\Phi(R,0))=g(K)$, where $K=\sqrt{1+\frac{p^2}{c^2}}-\frac{1}{\eta}c\cdot p$ and $\Phi(R,0)=0$ (if $\nabla\Phi(r,0)|_{r=R}=0$). Because of certain relations between p and K, once the expansion coefficient c_i in $f(R,p,t=0)=\sum_i c_ip^i$ is known, the expansion coefficient d_i in $g(K)=\sum_i d_iK^i$ are also easy to be calculated.

Clearly, BGK modes [23] are analytic exact solutions of the VE in $B\equiv 0$ case:

$$0=\left[\partial_t - E\left(r-\frac{1}{\eta}ct\right)\cdot\partial_{p(v)}\right]f+v\cdot\nabla f, \tag{20}$$

whose solutions are monovariable functions of $\phi\left(r-\frac{1}{\eta}ct\right)+\sqrt{1+p^2}-\frac{1}{\eta}c\cdot p$, where ϕ is scalar potential and $E=-\nabla\phi$. Moreover, there is a similar procedure of determining function form of g.

We should note that K is a nonlinear function of v and the maximum value of K or $K_{\max}$ is reached at $v=\frac{1}{\eta}c$. Thus, even if g is a Dirac function of $K+\Phi$, g cannot be a Dirac function of v, i.e., $g\sim\delta(v-u(r,t))$. The nonlinear function relation between K and v determines that g is at least a summation of two Dirac components: $g=f_1(r,t)\delta(v-u_1(r,t))+f_2(r,t)\delta(v-u_2(r,t))+\ldots$. This agrees with previous conclusion that functions of a general form $F_1(r,t)\delta(v-F_2(r,t))$ cannot meet the VE.

We should also note that, because of nonlinear function relation between g and $K+\Phi$, the maximum of g, or $g_{\max}$, is usually reached at $K+\Phi\neq K_{\max}+\Phi$. Namely, if $g_{\max}$ is reached at $K=K_{g\max}$, this $K_{g\max}$ usually corresponds to two values of v. In contrast, $K_{\max}$ merely corresponds to a value of v. Thus, the contour plot of g in the phase space will take on complicated structures, such as hole, island, etc. [17, 19].

5. Summary

By analyzing the bottleneck of numerical study the 6D phase space dynamics of a V-M system, we present, based on two universal properties, a universal and practical scheme of exact dynamics calculation of any V-M system. Our scheme is to harness, to the largest extent, available computer group condition to obtain as exact as possible solution. It gets rid of some inherent factors (of other methods), which yield incorrect physical results.

Author details

Hai Lin

Address all correspondence to: linhai@siom.ac.cn

State Key Laboratory of High Field Laser Physics, Shanghai Institute of Optics and Fine Mechanics, Shanghai, China

References

[1] Humpheries S. Charged Particle Beams. New York: Dover Pubs Inc; 2013

[2] Chaw AW. Physics of Beam Collective Effects in High Energy Accelerators. New York: Wiley; 1993

[3] Barnard JJ, Lund SM. Intense Beam Physics. New York: McGraw-Hill; 1995

[4] Moller SP. ELISA, and electrostatic storage ring for atomic physics. Nuclear Instruments and Methods in Physics Research Section A. 1997;**394**:281

[5] Tanabe T, et al. An electrostatic storage ring for atomic and molecular science. Nuclear Instruments and Methods in Physics Research Section A. 2002;**482**:595

[6] Andersen LH, Heber O, Zajfman D. Physics with electrostatic rings and traps. Journal of Physics B. 2004;**37**:R57

[7] Tessler DR, et al. Passive Electrostatic Recycling Spectrometer of Desk-Top Size for Charged Particles of Low Kinetic Energy. Physical Review Letters. 2007;**99**:253201

[8] Spanjers TL, et al. Evidences for isochronous behavior in electron and ion storage for a low energy electrostatic storage ring. Nuclear Instruments and Methods in Physics Research Section A. 2014;**736**:118

[9] Maier R, et al. The high-energy storage ring (HESR). In: Proceedings of the 24th Particle Accelerator Conference in Washington, DC, 2011. p. 2104

[10] Davidson RC, Chan HW, Chen CP, Lund S. Equilibrium and stability properties of intense non-neutral electron flow. Equilibrium and stability properties of intense non-neutral electron flow. Reviews of Modern Physics. 1991;**63**:341

[11] Chao AW, Pitthan R, Tajima T, Yeremian D. Physical Review Special Topics—Accelerators and Beams. 2003;**6**. DOI: 024201

[12] Vlasov A. On the Kinetic Theory of an Assembly of Particles with Collective Interaction. Journal of Physics (USSR). 1945;**25**:10

[13] Krall NA, Trivelpiece AW. Principles of Plasma Physics. New York: McGraw-Hill; 1977

[14] Kalman G. Nonlinear oscillations and nonstationary flow in a zero temperature plasma: Part I. Initial and boundary value problems. Annals of Physics 1960;**10**:1; Bertrand P, Feix MR. Non linear electron plasma oscillation: the water bag model. Physics Letters A 1968;**28**:68

[15] Dawson JM. Particle simulation of plasmas. Reviews of Modern Physics. 1983;**55**:403

[16] Birdsall CK, Langdon AB. Plasma Physics via Computer Simulation. New York: Taylor & Francis Group; 2004

[17] Lin H, Liu CP, Wang C, Shen BF. Macroscopic and Microscopic Structure of Electromagnetic Wakefield. The Open Plasma Physics Journal. 2014;**7**(1)

[18] Lin H, Liu CP, Wang C, Shen BF. Miniaturization of electron storage device. Europhysics Letters. 2015;**111**:62001

[19] Lin H. Non-Akhiezer-Polovin Model on Plasma Electrostatic Wave and Electron Beam. The Open Plasma Physics Journal. 2015;**8**(23)

[20] Lin H, Shen BF, Xu ZZ. Phase space coherent structure of charged particles system. Physics of Plasmas. 2011;**18**:062107

[21] Lin H. Exact solutions of macroscopic self-consistent electromagnetic fields and microscopic distribution of Vlasov-Maxwell system. arxiv.org e-print 1402.7072

[22] Katsouleas TC. Physical mechanisms in the plasma wake-field accelerator. Physical Review A. 1986;**33**:2056

[23] Bernstein IB, Greene JM, Kruskal MD. Exact Nonlinear Plasma Oscillations. Physical Review. 1957;**108**:546

Chapter 2

Motion of Electrons in Planar Ideal Undulator

Nikolay Smolyakov

Additional information is available at the end of the chapter

http://dx.doi.org/10.5772/intechopen.73682

Abstract

This chapter describes the motion of relativistic electrons in three-dimensional ideal undulator magnetic field. The undulator magnetic field satisfies the stationary Maxwell equations. Usually, the differential equations of electron motion in three-dimensional sinusoidal magnetic field are analysed by averaging over the fast electron oscillations. This averaging method was applied in a number of previously published papers. In this study, the nonlinear differential equations for electron motion were solved analytically by using the perturbation theory. The analytic expressions for trajectories obtained by this method describe the electron trajectories more accurately as compared with the formulas, which were obtained within the framework of the averaging method. An analysis of these expressions shows that the behaviour of electrons in such a three-dimensional field of the undulator is much more complicated than it follows from the equations obtained by the averaging method. In particular, it turns out that the electron trajectories in a planar undulator are cross-dependent. A comparison of the trajectories, calculated using these new analytical expressions with the numerically calculated trajectories using the Runge-Kutta method, demonstrated their high accuracy.

Keywords: undulator, wiggler, beam dynamics, storage ring, light source

1. Introduction

Here, a theoretical analysis of electron motion in a planar undulator (or wiggler) with ideal three-dimensional magnetic field is carried out. In this case, the magnetic field on the undulator axis (Z-axis, see **Figure 1**) is directed strictly vertically upwards (Y-axis) and has a perfect sinusoidal dependence on the longitudinal coordinate Z. However, similarly as in the case of real planar undulators, the ideal magnetic field considered here is supposed to be nonuniform in the transverse plane, that is, in the XOY plane. In the case of standard geometry undulator (an undulator with a plane surface of poles), as shown in **Figure 1**, the amplitude of

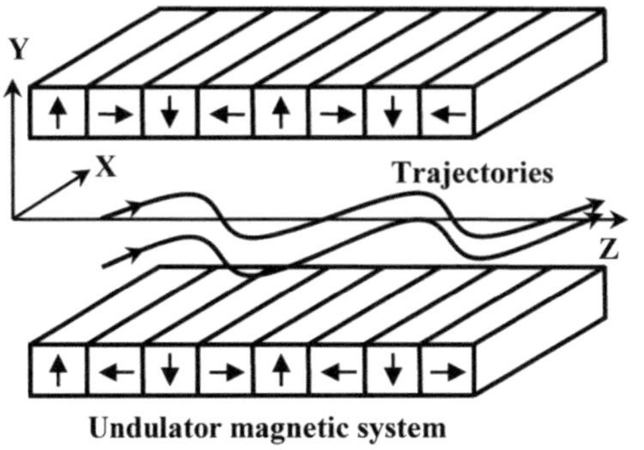

Figure 1. Scheme of electron motion in an undulator.

undulator magnetic field increases in the vertical direction on approaching its poles. As you move away from the axis in the undulator median plane in the horizontal direction (i.e. along *X*-axis), the amplitude of the vertical magnetic field decreases as you approach to the magnetic pole boundaries. The fact that the undulator magnetic field satisfies the stationary Maxwell equations imposes additional requirements on the functional dependence of the magnetic field components on the spatial coordinates. This also leads to the appearance of weak horizontal (along the *X*-axis) and longitudinal (along the *Z*-axis) components of the magnetic field in the region of the planar undulator median plane. Both these factors, that is, the presence of horizontal components of the magnetic field and the inhomogeneity of the field in the transverse plane produce the undulator focusing properties. This means that if two electrons enter the undulator magnetic field in parallel to each other though spaced apart from each other, then at the end of the undulator they will already have non-parallel velocities. This is because each of the electrons moves in its own, individual undulator magnetic field. A relativistic electron beam, in its passing through a planar undulator magnetic field, is focused in the vertical direction and is defocused in the horizontal direction, since the amplitude of the undulator field increases with distance from its axis in the vertical direction and, vice versa, decreases with distance from the axis of the undulator in the horizontal direction. These focusing and defocusing properties of the undulator magnetic field have a strong influence on the electron beam dynamics in the electron storage rings, since the undulator in this respect manifests itself as an additional quadrupole lens. This leads to a shift of radial and vertical betatron oscillation frequencies of electron beam in the electron storage ring and, respectively, to the displacement of its working point. It can dramatically decrease the electron beam lifetime, since the displaced working point may fall into the resonance region. Thus, accurate consideration of the undulator focusing properties is of great importance for understanding the electron beam dynamics in the electron storage rings.

As far as we know, the focusing properties of the planar undulator magnetic field were first theoretically predicted in [1], where the effects of the superconducting wiggler influence on the storage ring electron beam dynamics were analysed. The horizontal and vertical focal lengths were also calculated. It has been shown that these focusing properties have a detectable effect on the electron beam dynamics. For example, they shift vertical and horizontal betatron oscillation frequencies of the electron beam. Assuming a planar undulator with infinitely wide poles, its magnetic field is uniform along the horizontal X-axis (**Figure 1**). It is evident that there is no horizontal focusing in this case, and such undulator focuses the electron beam vertically only. The motion of electrons in a planar undulator with plane infinitely wide poles was analysed in papers [2, 3]. A clear physical explanation for vertical focusing effect in a two-dimensional magnetic field of such undulator was also given in [2]. As a consequence of Maxwell equations, the undulator magnetic field outside its median plane also has a longitudinal component, directed along the undulator Z-axis. This longitudinal component has an alternating (sinusoidal) character, that is, it is either aligned with the Z-axis or opposing the undulator axis. The phase of this component is determined by the phase of the leading (vertical) undulator magnetic field. Likewise, this leading vertical field causes the electron to oscillate in the horizontal plane, resulting in the horizontal (along the X-axis) sinusoidal component of the electron velocity. The phase of this horizontal component of the electron velocity is also determined by the phase of the undulator leading field. The action of this longitudinal sinusoidal component of the undulator magnetic field on an electron, which proceeds along the horizontally oscillating trajectory, leads to the relatively small vertically directed Lorentz force. The mutual correlation of the longitudinal component of the undulator magnetic field (directed along the undulator axis) and the horizontal component of the electron velocity (directed along the X-axis) are such that the Lorentz force is always directed towards the median plane of the undulator, thus creating vertical focusing force [2].

Some general relationships between the vertical and horizontal focal lengths of the undulator were derived in papers [4–8]. The general expressions for calculating horizontal and vertical focal lengths are also derived in the case of an undulator with flat finite-width poles, which are alternately shifted in the horizontal direction (along X-axis) relative to each other (undulators with the poles offset) [4, 5]. In the standard case of the undulator with zero-offset geometry, these formulas transform into the corresponding expressions given in paper [1]. Electron long-wave anharmonic betatron oscillations in very long undulator magnetic fields were considered in [9]. The action of the focusing properties of undulators on the operation of free-electron lasers was studied in [10–12]. In addition, a configuration of an undulator with a parabolic shape of the magnetic-pole surface was also proposed in [10]. Such geometry of a magnetic-pole surface leads to a rise in amplitude of the undulator magnetic field as the distance from the undulator axis in the median plane increased. As a result, both horizontal focal length and vertical focal length became positive. Therefore, this undulator focuses the electron beam in both directions, which is important for the free-electron lasers operation. The papers [13, 14] considered the influence of the inhomogeneities in the transverse plane of the electromagnetic wave and helical magnetic field of spiral undulator on the generation of X-ray radiation in free-electron lasers. The following mathematical method was used in all above-mentioned papers. The focal lengths of the undulator were calculated in the framework of a smoothed (focusing) approximation. The averaging procedure of the electron trajectory in the undulator sinusoidal magnetic field over the oscillation period plays an important role in such kind of calculation. The following section describes this approximation in more detail. It is generally accepted that this procedure of oscillation averaging is correct and corresponds to the physics of the process. If the influence of the undulator magnetic field on the electron beam dynamics is reduced in the undulator focusing properties, then it necessarily implies averaging over oscillations of the sinusoidal-type electron trajectory in the sign-changing undulator magnetic field. Paper [15] was one of the first papers that described this approach that was applied for the analysis of influence of nonuniformities of the undulators and wigglers magnetic fields on the electron beam dynamic. In succeeding years, it was extensively used in studies involving the influence of wigglers and undulators on the electron beam dynamic in electron storage rings [16–20].

The wavelength λ of the fundamental (first) harmonic along the undulator axis is $\lambda = \frac{\lambda_u}{2\gamma^2}(1 + \frac{1}{2}K^2)$. Here, λ_u is an undulator period, γ is the electron-reduced energy, and K is the undulator deflection parameter. In a free-electron laser, the value of the dimensionless Pierce parameter ρ determines the width of the spectral line of the coherent electromagnetic radiation. This means that the width of the spectral line $\Delta\lambda$ of the fundamental harmonic must be known with accuracy determined by the condition $\Delta\lambda/\lambda \leq \rho$. In the case of the European XFEL facility (Hamburg), we have $\gamma \cong 35,000$, $K \cong 4$, and the value of the Pierce parameter is equal to $3{\cdot}10^{-4}$ for a radiation wavelength of 0.1 nm [21]. Some variations in magnetic field strength are manifested in correspondent changes of effective undulator deflection parameter K Variation of wavelength λ of the undulator radiation fundamental harmonic due to variation of the undulator deflection parameter is approximately equal to $\Delta\lambda/\lambda \cong 2\Delta K/K$. Therefore, the absolute accuracy, with which the undulator deflection parameter K should be calculated, must satisfy the following condition: $\Delta K < 0.5K\rho$. It follows from the general formula for the phase of spontaneous radiation that the deflection parameter K of planar undulator with periodic magnetic field is determined by the following relation: $K^2 = 2\frac{\gamma^2}{\lambda_u}\int_0^{\lambda_u} \beta_x^2(z)dz$, where β_x is the horizontal component of the reduced electron velocity, which has a sinusoidal form with the amplitude approximately equal to K/γ. Therefore, calculations of the transverse component of the electron velocity must be carried out with very high accuracy, which is determined by the relation $\Delta\beta_x = \Delta K/\gamma < (K\rho)/(2\gamma) \approx 2 \cdot 10^{-0}$. The relative accuracy in this case is of the order of $2{\cdot}10^{-4}$. These requirements for the simulation accuracy show that the abovementioned focusing approximation, based on the method of averaging fast oscillations of the electron trajectory, in some cases may not have the sufficient accuracy. It is clear that for free-electron laser, we need to develop a more precise method for calculations of the electron trajectories in planar undulator three-dimensional magnetic fields.

In a number of recent papers [22–24], electron trajectories in perfect sinusoidal three-dimensional magnetic field of a planar undulator were numerically simulated. The Runge-Kutta algorithm was employed for solving the set of differential equations for electron motion in the undulator field. It is correct to suppose that these numerically simulated trajectories are highly accurate results. The checking of these numerically simulated trajectories was made against analytically calculated trajectories, obtained by using the oscillation-averaging method (focusing approximation) [10–12]. This comparison has been demonstrated in a conclusive way that in most cases the numerically simulated trajectories differ significantly from those calculated by using the analytical formulas derived in focusing approximation. Therefore, more precise analytical formulas for electron trajectories in an ideal sinusoidal three-dimensional

http://dx.doi.org/10.5772/intechopen.73682

magnetic field of a planar undulator are critically important to properly understand electron beam dynamics.

Here, we derive new analytical expressions for trajectories of relativistic electrons in the ideal three-dimensional magnetic field of a planar undulator (or a wiggler). It means that the undulator magnetic field has only the vertical component at the undulator axis with pure sinusoidal form. However, outside the undulator axis, there are the horizontal and longitudinal components of the magnetic field. All three components of the magnetic field are related to each other functionally since the undulator magnetic field must satisfy the stationary Maxwell equations. The differential equations of motion for electrons in such a magnetic field were solved by using the perturbation theory, which is widely used in quantum mechanics rather than the focusing approximation which employs the averaging over transverse oscillations of the electron trajectory. The idea of this method for trajectory calculating was suggested in paper [25] for the first time. The formulas derived in this manner are very complicated since they include all terms of the cubic power of small quantities. However, these formulas give a higher approximation to electron trajectories in the undulator field than those derived in the smoothed (focusing) approximation [10–12]. Analysis of these highly accurate expressions shows that electron motion in undulator magnetic field is very sophisticated and cannot be reduced to the standard focusing effects. In particular, the electron motion in the vertical and horizontal directions is interrelated. This means that the change in the initial conditions of electrons in the vertical plane results in the correspondent changes of the horizontal component of the electron trajectory and vice versa. It is reasonable because the Maxwell equations for the stationary magnetic field interrelate all three components of the undulator field. However, this effect cannot be described within the framework of the smoothed (focusing) approximations.

Using the Runge-Kutta algorithm, a computer code was used to numerically solve differential equations for motion of an electron in the three-dimensional planar undulator magnetic field. Comparison of the numerically calculated trajectories with those derived from the analytical accurate formulas demonstrates a very high accuracy of these analytical expressions. However, it is clear that, in practical use, the analytical expressions are often vastly superior to numerical simulations., A step-by-step calculation with a small interval along all trajectories is required for purposes of the electron trajectory numerical simulations. This procedure takes a good deal of time. In the event that we know the highly accurate analytical expressions for describing electron trajectories in the planar undulator, we can calculate the final coordinates and velocity of the electron easily by simply substituting the final value of the magnetic field longitudinal coordinate into the analytical expressions. This greatly reduces the computation time.

2. Equations of electron motion in ideal planar undulator

Let $\vec{B}(x, y, z)$ be a planar undulator or wiggler magnetic field produced by its magnetic system. The equation of electron motion in this field has the form:

$$\frac{d\vec{\beta}\,(t)}{dt}=\frac{e}{mc\gamma}\left[\vec{\beta}(t)\times\,\vec{B}(x(t),y(t),z(t))\right] \tag{1}$$

where e, m, $\vec{\beta}$ and γ are the electron charge, mass, reduced velocity and reduced energy, respectively; $e<0$ and $\vec{r}\,(t)=\{x(t),y(t),z(t),\}$ are the electron trajectories. Let us recall that the electron energy and velocity modulus are constant in the magnetic field: $\gamma=const$, $\beta=const$.

The time t is an independent variable in equations of motion (1). At the same time, the undulator magnetic field in Eq. (1) is a function of the transversal spatial coordinates x, y and longitudinal coordinate z. Consequently, it is more convenient to use the new independent variable z (the longitudinal coordinate) instead of the independent variable t (time). With the equations of motion (1), it is possible to derive the following exact equations for the electron trajectory in an external magnetic field [26, 27]:

$$x''=-\left(e/\left(mc^2\beta\gamma\right)\right)\sqrt{1+(x')^2+(y')^2}\left[\left(1+(x')^2\right)B_y-y'B_z-x'y'B_x\right] \tag{2}$$

$$y''=\left(e/\left(mc^2\beta\gamma\right)\right)\sqrt{1+(x')^2+(y')^2}\left[\left(1+(y')^2\right)B_x-x'B_z-x'y'B_y\right] \tag{3}$$

We point out that the Eqs. (2), (3) are expressed in terms of the longitudinal coordinate z. The prime in the Eqs. (2), (3) means differentiation with respect to z.

Here, the undulator with planar magnetic system and ideal three-dimensional sinusoidal magnetic field is considered; see **Figure 1**:

$$B_x(x,y,z)=-\left(k_x/k_y\right)B_0\sin(k_xx)\sinh\left(k_yy\right)\sin(k_zz) \tag{4}$$

$$B_y(x,y,z)=B_0\cos(k_xx)\cosh\left(k_yy\right)\sin(k_zz) \tag{5}$$

$$B_z(x,y,z)=\left(k_z/k_y\right)B_0\cos(k_xx)\sinh\left(k_yy\right)\cos(k_zz) \tag{6}$$

where B_0 is the magnetic field amplitude on the undulator axis (Z-axis), λ_u is the undulator period length, $k_x=1/a$, $k_z=2\pi/\lambda_u$ and $k_y=\sqrt{k_x^2+k_z^2}$.

The parameter a determines the magnetic field nonuniformity along the horizontal X-axis. It is of the order of the undulator pole width. In the case of undulator with infinitely wide poles, the parameter $a=\infty$ and $k_x=0$. It is easy to verify that the magnetic field, which is described by Eqs. (4)–(6), satisfies the Maxwell equations for a stationary magnetic field.

The system of precise Eqs. (2), (3) for the electron motion appears as cumbersome formulas. Nevertheless, it offers several advantages in analytical analysis and numerical simulations over the standard equations of motion (1). First, the undulator magnetic field is described by using the functions of the longitudinal coordinate z. Consequently, in this case, the functions $\sin(k_zz)$ and $\cos(k_zz)$ in Eqs. (4)–(6) are known exactly. When employing the standard equations of motion (1), the value of the electron's longitudinal coordinate $z(t)$ at every step is calculated with some finite precision and the resultant errors are accumulated. It is also significant that the system of Eqs. (1) includes three equations, while the system of Eqs. (2), (3) consists of two equations only. This also simplifies its analysis and yields a large dividend in accuracy.

The region occupied by the electron beam, that is, the small vicinity near the undulator axis, has relatively small transversal coordinates: $|k_x x| \ll 1$, $|k_y y| \ll 1$. Expanding Eqs. (4)–(6) in terms of these small quantities, we have the following expressions:

$$B_x(x, y, z) \cong -B_0 k_x^2 xy\sin(k_z z) \tag{7}$$

$$B_y(x, y, z) \cong B_0\left(1n0.5k_x^2 x^2 + 0.5k_y^2 y^2\right)\sin(k_z z) \tag{8}$$

$$B_z(x, y, z) \cong B_0 k_z y\cos(k_z z) \tag{9}$$

It is clear that on the undulator axis $x = 0$ and $y = 0$ only when vertical component of the magnetic field is nonzero: $B_x(0, 0, z) = 0$, $B_z(0, 0, z) = 0$, $B_y(0, 0, z) = B_0\sin(k_z z)$. In this regard, the field is ideal because the magnetic field of the real undulator inevitably includes errors caused by manufacturing errors of the undulator magnetic system. There are also higher harmonics in the magnetic field generated by the real magnetic system. Their relative amplitudes depend on specific details of the undulator design.

Differential Eqs. (2), (3) are nonlinear and cannot be solved exactly. However, the functions x', y', $k_x x$ and $k_y y$ in Eqs. (2), (3) are small in absolute value. Therefore, we can expand the nonlinear differential Eqs. (2), (3) in terms of these small quantities. Substituting the expressions (4–6) for the undulator magnetic field, as a result we have:

$$x^{''} = \widetilde{K}k_z\left\{\left(1 + 0.5\left(3(x')^2 - k_x^2 x^2 + (y')^2 + k_y^2 y^2\right)\right)\sin\varphi - k_z yy'\cos\varphi\right\} \tag{10}$$

$$y^{''} = \widetilde{K}k_z\left(\left(k_x^2 xy + x'y'\right)\sin\varphi + k_z x'y\cos\varphi\right) \tag{11}$$

where $K = \frac{-eB_0\lambda_u}{2\pi mc^2}$ is the undulator deflection parameter, $e < 0$ for electrons, $\varphi = k_z z$, and $\widetilde{K} = \frac{K}{\beta\gamma}$.

In the cases of our interest, the dimensionless undulator deflection parameter K is of the order of several units, that is, ~1–5, and the reduced electron energy γ is of the order of several thousands. So, $\gamma \cong 5000$ for the electron beam of the Sibiria-2 electron storage ring (Moscow) and $\gamma \cong 35,000$ for the European XFEL facility (Hamburg). Because of this, $\widetilde{K}$ is much less than unity: $\widetilde{K} \sim 10^{-3} - 10^{-4}$.

Neglecting all small terms x', y', $k_x x$ and $k_y y$ in Eqs. (10), (11), we get the following equations in linear approximation:

$$x^{''} = \widetilde{K}k_z\sin\varphi \tag{12}$$

$$y^{''} = 0 \tag{13}$$

The solutions of Eqs. (12), (13) are the following:

$$x_1(z) = x_0 + \theta_0 z - \left(\widetilde{K}/k_z\right)\sin\varphi \tag{14}$$

$$y_1(z) = y_0 + y'_0 z \tag{15}$$

where $\theta_0 = x'_0 + \widetilde{K}$ is the initial deviation of the electron velocity from its equilibrium value.

Eqs. (14), (15) correspond to rectilinear electron motion with additional sinusoidal oscillations in the horizontal plane. Obviously, Eqs. (14), (15) do not describe any focusing properties of the undulator magnetic field. They describe motion of an electron in the magnetic field with the following parameters: $B_x = 0$, $B_y(z) = B_0\sin(k_z z)$ and $B_z = 0$. Clearly, these formulas describe the magnetic field at the undulator axis. To put it differently, they follow from more general Eqs. (4)–(6) if we neglect the magnetic field's nonuniformity in the transverse plane and the small magnetic field components B_x and B_z. This primitive magnetic field cannot be produced by real magnetic system because it does not satisfy the Maxwell equations. At the same time, it is often employed when characteristics of electromagnetic radiation from a planar undulator are analysed.

3. Smoothed (focusing) approximation for electron trajectories

It is reasonable to generalise the Eqs. (14), (15) as follows. We replace the terms $x_0 + \theta_0 z$ and $y_0 + y'_0 z$ in Eqs. (14), (15) by slowly varying functions of the general form $x_s(z)$ and $y_s(z)$, respectively. So, we seek the electron trajectory in the following form:

$$x_f(z) = x_s(z) - \left(\widetilde{K}/k_z\right)\sin\varphi \tag{16}$$

$$y_f(z) = y_s(z) \tag{17}$$

We substitute Eqs. (16), (17) into equations of motion (10, 11) and average them over the undulator period. It makes all fast oscillating terms equal to zero. This means that odd powers of the functions $\sin\varphi$ and $\cos\varphi$ vanish, and the functions $\sin^2\varphi$ and $\cos^2\varphi$ are replaced by their average value 0.5. As a consequence, we obtain the following linear differential equations for the slowly varying functions $x_s(z)$ and $y_s(z)$, see [10–12]:

$$x''_s(z) - k_z^2\omega_x^2 x_s(z) = 0 \tag{18}$$

$$y''_s(z) + k_z^2\omega_y^2 y_s(z) = 0 \tag{19}$$

where $\omega_{x,y} = \widetilde{K}k_{x,y}/\left(\sqrt{2}k_z\right)$ are the dimensionless frequencies of betatron oscillations in the horizontal and vertical directions in units of the undulator period λ_u, respectively. The quantities $\omega_{x,y}$ have the same order in magnitude as the parameter $\widetilde{K}$ and are also small. The quantity a is of the order of the pole width. Usually, the pole width is slightly larger than the undulator period length. Therefore, the following condition is usually true for undulators: $k_x/k_z = \lambda_u/$

$(2\pi a) \leq 1$. It is clear that $k_y = \sqrt{k_x^2 + k_z^2} > k_x$, and hence $\omega_x < \omega_y$. In the case of the undulator with infinitely wide poles, we have $a = \infty$, $k_x = 0$ and $\omega_x = 0$.

Since the functions $x_s(z)$ and $y_s(z)$ are solutions of the Eqs. (18), (19), they consist of the linear combinations of hyperbolic sines and cosines (for Eq. (18)) and trigonometric sines and cosines (for Eq. (19) correspondently). Taking Eqs. (16), (17) into account, thus we have in the smoothed (focusing) approximation the following expressions for the horizontal and vertical coordinates of the electron in the magnetic field given by expressions (4–6):

$$x_f(z) = x_0 \cosh(\omega_x \varphi) + \left((\theta_0/(\omega_x k_z))\sinh(\omega_x \varphi) - \left(\widetilde{K}/k_z\right)\sin\varphi \right. \tag{20}$$

$$y_f(z) = y_0 \cos(\omega_y \varphi) + \left(y_0'/(\omega_y k_z)\right)\sin(\omega_y \varphi) \tag{21}$$

$$x_f'(z) = x_0 \widetilde{K}\left(k_x/\sqrt{2}\right)\sinh(\omega_x \varphi) + \theta_0 \cosh(\omega_x \varphi) - \widetilde{K}\cos\varphi \tag{22}$$

$$y_f'(z) = -y_0 \widetilde{K}\left(k_y/\sqrt{2}\right)\sin(\omega_y \varphi) + y_0' \cos(\omega_y \varphi) \tag{23}$$

It is significant that Eqs. (20)–(23) are linear in terms of the initial electron parameters x_0, y_0, θ_0 and y_0'. For the undulator with infinitely wide poles, we have $\omega_x = 0$, and Eq. (20) for the horizontal component of the electron trajectory coincides with Eq. (14).

Two linear equations for electron motion (18, 19) are decoupled in the smoothed (focusing) approximation. It implies that the first Eq. (18) is dependent only on the parameters of the horizontal component of trajectory. Correspondingly, the second Eq. (19) is dependent on the vertical component parameters. In other words, these both equations of motion are completely independent of each other. Respectively, the Eqs. (20), (21) are also decoupled, that is, they are independent of each other. However, the more precise system of equations of motion (10, 11) is not decoupled. It means that each of these equations depends explicitly on the parameters of the horizontal and vertical alike components of the electron trajectory. As a result, every component of the electron trajectory, both horizontal and vertical, being the solutions of the system of equations (10, 11), must also be dependent on both horizontal and vertical parameters of electron trajectory.

4. Trajectories in a short undulator

Magnetic fields of short planar undulators have focusing properties, that is, the influence of short undulator magnetic field on the electron beam dynamics can be described in terms of the undulator focal lengths. However, the ideal magnetic field deflects the propagating electron beam in the median plane. As a result, there is no straight electron trajectory (principal axis) in an undulator. The absence of axial symmetry leads to astigmatism, that is, the undulator horizontal and vertical focal lengths are different and even have different signs. The vertical focal length is positive, while the horizontal focal length is negative.

We consider relatively short undulators with the number of periods N so that the following condition is fulfilled:

$$\sqrt{2}\pi NK/(\gamma\beta) \ll 1 \tag{24}$$

In the cases under consideration: $\widetilde{K} = K/(\gamma\beta) \sim 10^{-1} - 10^{-0}$ and standard number of undulator periods is about $N \sim 100$. Therefore, the inequality (24) almost without exception is fulfilled. Since $\varphi = 2\pi z/\lambda_u \leq 2\pi N$, the quantity φ increases linearly along the undulator length with the maximum value equal to $2\pi N$. Quantities $\omega_{x,y}$ have the same order as $\widetilde{K}$. It follows from inequality (24) that the conditions $\omega_{x,y}\varphi \ll 1$ are always true for any point of the electron trajectory in a short undulator. Therefore, we can expand Eqs. (20)–(23) in terms of these small quantities $\omega_{x,y}\varphi$ and retain terms to powers less or equal than 3. As a result, we have in terms of z:

$$x_f(z) = x_0 + \theta_0 z - \left(\widetilde{K}/k_z\right)\sin(k_z z) + 0.25x_0\widetilde{K}^2(z/a)^2 + (\theta_0/12)\widetilde{K}^2\left(z^3/a^2\right) \tag{25}$$

$$y_f(z) = y_0 + y_0' z - 0.25y_0\widetilde{K}^2\left(k_y z\right)^2 - \left(y_0'/12\right)\widetilde{K}^2 k_y^2 z^3 \tag{26}$$

$$x_f'(z) = \theta_0 - \widetilde{K}\cos(k_z z) + 0.5x_0\widetilde{K}^2 z/a^2 + 0.25\theta_0\widetilde{K}^2(z/a)^2 \tag{27}$$

$$y_f'(z) = y_0' - 0.5y_0\widetilde{K}^2 k_y^2 z - 0.25\cdot y_0'\widetilde{K}^2 k_y^2 z^2 \tag{28}$$

Eqs. (25)–(28) determine the electron trajectory in a short undulator which is defined by the inequality (24). Let us compare Eqs. (25), (26) which are derived in the framework of focusing approximation, with Eqs. (14), (15) obtained in the linear approximation. It is clear that all additional terms describing the focusing properties of the undulator have the cubic power for the small parameters $\widetilde{K}$, x_0, y_0, θ_0 and y_0', namely $x_0\widetilde{K}^2$, $\theta_0\widetilde{K}^2$, $y_0\widetilde{K}^2$ and $y_0'\widetilde{K}^2$.

Since we know the electron trajectories in the short undulator (which is specified by the inequality (24)), we can calculate its focal lengths. We first consider an electron moving along the equilibrium trajectory. This trajectory is defined by the following initial conditions $x_0 = y_0 = \theta_0 = y_0' = 0$ and is described by the formulas:

$$x_{eq}(z) = -\left(\widetilde{K}/k_z\right)\sin(k_z z) \tag{29}$$

$$x_{eq}'(z) = -\widetilde{K}\cos(k_z z) \tag{30}$$

$$y_{eq}(z) = y_{eq}'(z) = 0 \tag{31}$$

Let us consider another electron, which enters the undulator in parallel to the first one but is shifted upward, that is, its initial conditions are $x_0 = \theta_0 = y_0' = 0$, $y_0 > 0$. Its trajectory is defined by the formulas:

$$x_2(z) = x_{eq}(z) = -\left(\widetilde{K}/k_z\right)\sin(k_z z) \tag{32}$$

$$x_2'(z) = x_{eq}'(z) = -\widetilde{K}\cos(k_z z) \tag{33}$$

$$y_2(z) = y_0 - 0.25 y_0 \widetilde{K}^2 (k_y z)^2 \tag{34}$$

$$y_2'(z) = -0.5 y_0 \widetilde{K}^2 k_y^2 z \tag{35}$$

At the undulator end with the Z-coordinate $z_N = N\lambda_u$, the electron has the following vertical coordinate and velocity:

$$y_2(z_N) = y_0\left[1 - \left(\pi N\widetilde{K}\right)^2\left(1 + \lambda_u^2/(2\pi a)^2\right)\right] \tag{36}$$

$$y_2'(z_N) = -y_0 \cdot 2\pi^2\widetilde{K}^2 N\left[1 + \lambda_u^2/(2\pi a)^2\right]/\lambda_u \tag{37}$$

Taking into account inequality (24) for short undulators, it is easy to see from Eq. (36) that we can neglect by the vertical shift of the electron inside the undulator: $y_2(z_N) \cong y_0$. It is also clear that, after exiting the undulator, two electrons with the trajectories described by Eqs. (29)–(31) and (32)–(35), respectively, intersect each other at the vertical undulator focus with the mutual angle $y_2(z_N)/f_y \cong y_0/f_y$, where f_y is the vertical focal length of the undulator. On the other hand, the intersection angle is equal to $-y_2'(z_N)$. As a result, with the help of Eq. (37), we get the expression for vertical focal length f_y:

$$\frac{1}{f_y} = \frac{2\pi^2 K^2 N}{\lambda_u \gamma^2}\left[1 + \lambda_u^2/(2\pi a)^2\right] \tag{38}$$

Similarly, it is easy to derive the expression for the horizontal focal length:

$$\frac{1}{f_x} = -\frac{K^2 N\lambda_u}{2a^2\gamma^2} \tag{39}$$

By applying slightly other methods, the expressions for vertical and horizontal focal lengths (38) and (39) were derived in the previous works [1, 4–8].

The foregoing shows that the solutions of the equations for electron motion in the ideal magnetic field of a short undulator, obtained with employing method of the averaging of trajectory of fast oscillations include the focusing properties of the magnetic field. That is why the smoothed approximation can also be called as focusing. The Eqs. (38), (39) show that the vertical focal length is positive (the electron beam is focused in vertical direction), while the horizontal focal length is negative (electron beam is defocused in horizontal direction). The focusing powers of the undulator (the quantities inverse to the focal lengths) $1/f_{x,y}$ are proportional to the number of undulator periods N and to the squared undulator deflection parameter K^2 and are inversely proportional to the squared electron beam energy γ^2.

By using Eqs. (38), (39), it is easy to derive the following general relation:

$$\frac{1}{f_x} + \frac{1}{f_y} = \frac{2\pi^2 K^2 N}{\lambda_u \gamma^2} \tag{40}$$

The key feature of Eq. (40) is that it is independent, which determines the value of the magnetic-field decay of the magnetic field (see Eqs. (4)–6)) along the horizontal axis X. It is clear that in the case of infinitely wide magnetic poles, that is, at $a = \infty$, the horizontal focal length also tends to infinity: $f_x = \infty$.

5. Electron trajectory calculation by methods of perturbation theory

It is possible to enhance considerably the accuracy of the solution to the equations of motion (20)–(23) as follows: Let us try to find the solution to Eqs. (10) and (11) in the form:

$$x(z) = x_f(z) + \Delta x(z) \tag{41}$$

$$y(z) = y_f(z) + \Delta y(z) \tag{42}$$

We assume that the unknown functions $\Delta x(z)$ and $\Delta y(z)$ are far less than the leading terms $x_f(z)$ and $y_f(z)$. We substitute Eqs. (41), (42) into Eqs. (10), (11) and ignore the functions $\Delta x(z)$ and $\Delta y(z)$ on the right-hand side of these two equations. As a result, we get two second-order differential equations, whose right-hand sides are well defined and expressed in terms of the products of trigonometric and hyperbolic functions:

$$(\Delta x)'' + x_f'' = \widetilde{K} k_z \left\{ \left(1 + 0.5 \left(3 \left(x_f' \right)^2 - k_x^2 x_f^2 + \left(y_f' \right)^2 + k_y^2 y_f^2 \right) \right) \sin\varphi - k_z y_f y_f' \cos\varphi \right\} \tag{43}$$

$$(\Delta y)'' + y_f'' = \widetilde{K} k_z \left(\left(k_x^2 x_f y_f + x_f' y_f' \right) \sin\varphi + k_z x_f' y_f \cos\varphi \right) \tag{44}$$

The functions x_f, x_f', x_f'', y_f, y_f' and y_f'' are well known and are expressed in elementary function by Eqs. (20), (21). The unknown functions $\Delta x(z)$ and $\Delta y(z)$ can be found by double integration of Eqs. (46), (47) over the variable z. For simplicity, we consider here the case of a short undulator: $\sqrt{2}\pi NK/(\gamma\beta) \ll 1$ where N is the number of undulator periods. As a result, omitting technically cumbersome intermediate calculations, we arrive at the following extremely complicated expressions:

$$\tilde{x}(z) = \tilde{x}_0\cosh(\omega_x\varphi) + (\theta_0/\omega_x)\sinh(\omega_x\varphi) - \tilde{K}[1 - 0.5k_x^2(x_0 + \theta_0 z)^2 + 0.5k_y^2(y_0 + y_0' z)^2]\cdot\sin\varphi - (0.5\tilde{K}/A^2)\tilde{x}_0^2\varphi - (2\tilde{K}/A^2)\theta_0(\tilde{x}_0 + \theta_0\varphi)(1 - \cos\varphi) + 1.5\tilde{K}\theta_0^2(1 + 2/A^2)(\varphi - \sin\varphi) + (\omega_y^2/\tilde{K})\tilde{y}_0^2\varphi + \tilde{K}y_0'(\tilde{y}_0 + y_0'\varphi)(1 + 2/A^2)(1 - \cos\varphi) - 0.5\tilde{K}(y_0')^2(1 + 6/A^2)(\varphi - \sin\varphi) - 0.25(\tilde{K}^2/A^2)(\tilde{x}_0 + \theta_0\varphi)\sin^2(\varphi) - 0.125\tilde{K}^2\theta_0(3 - 1/A^2)\left(2\varphi - \sin(2\varphi)\right) + 0.375\tilde{K}^3(1 - 1/A^2)(\varphi - \sin\varphi) + (\tilde{K}^3/24)\left(1 + 1/(3A^2)\right)\left(3\varphi - \sin(3\varphi)\right) \quad (45)$$

$$\tilde{y}(z) = \tilde{y}_0\cos(\omega_y\varphi) + (y_0'/\omega_y)\sin(\omega_y\varphi) - \left(\tilde{K}/A^2\right)(\tilde{x}_0 + \theta_0\varphi)(\tilde{y}_0 + y_0'\varphi)\sin\varphi + \left(\tilde{K}/A^2\right)\tilde{x}_0\tilde{y}_0\varphi + \left[\left(2\tilde{K}/A^2\right)(\tilde{x}_0 + \theta_0\varphi)y_0' + \tilde{K}(1 + 2/A^2)\theta_0(\tilde{y}_0 + y_0'\varphi)\right](1 - \cos\varphi) - \tilde{K}\theta_0 y_0'(1 + 6/A^2)(\varphi - \sin\varphi) - 0.25\omega_x^2 y_0'(2\varphi - \sin(2\varphi)) - 0.25\tilde{K}^2(1 - 1/A^2)(\tilde{y}_0 + y_0'\varphi)\sin^2(\varphi) \quad (46)$$

where $\tilde{K} = K/(\beta\gamma)$, $A = k_z a$, $\tilde{x}(z) = k_z x(z)$, $\tilde{y}(z) = k_z y(z)$, $\tilde{x}_0 = k_z x_0$, $\theta_0 = x'_0 + \tilde{K}$, $\tilde{y}_0 = k_z y_0$, $\omega_x = \tilde{K}k_x/(\sqrt{2}k_z) = \tilde{K}/(\sqrt{2}A)$, $\omega_y = \tilde{K}k_y/(\sqrt{2}k_z) = \left(\tilde{K}/\sqrt{2}\right)\sqrt{1 + 1/A^2}$, $\varphi = k_z z$.

Eqs. (45), (46) completely determine the electron trajectory in three-dimensional undulator magnetic field, which is described by the Eqs. (4)–(6). Differentiating Eqs. (45), (46) with respect to the longitudinal coordinate z, we get the corresponding formulas for the transverse components of the electron-reduced velocity $\beta_x(z) \cong x'(z)$ and $\beta_y(z) \cong y'(z)$.

Eqs. (45), (46) include all terms that are linear and cubic in small values $\tilde{K}$, x_0, y_0, θ_0 and y'_0. That is why these formulas are so cumbersome. Some terms in Eqs. (45), (46) are quadratic in terms of the electron initial parameters x_0, y_0, θ_0 and y'_0. It is natural since the equations of motion (10, 11) are nonlinear in terms of functions $x(z)$ and $y(z)$.

The first three terms in Eq. (45) include Eq. (20) for trajectories, which were derived in the focusing approximation. However, the third term in brackets in Eq. (45) contains additional quadratic terms, which have a clear physical meaning. They correspond to a change in the undulator magnetic field amplitude during the electron motion along a straight line. This straight line is the electron trajectory averaged over fast horizontal oscillations. The first two terms of Eq. (46) coincide with Eq. (21) for the vertical component of the electron trajectory. We mention that formulas (45) and (46) are given in terms of reduced dimensionless coordinates and $\tilde{x}(z) = k_z x(z)$ and $\tilde{y}(z) = k_z y(z)$.

Some terms in Eqs. (45), (46) include the factor $\varphi = k_z z$, which linearly increases along the undulator with increasing of the longitudinal coordinate z. It has a maximum $2\pi N$ large in value at the final point of the undulator magnetic field at $z_N = \lambda_u N$. It is clear that such terms make large contributions to the expressions for the electron trajectory. It is also easy to see from Eqs. (45), (46) that the majority of cubic terms oscillate. These terms clearly vanish with the averaging over fast oscillation procedure. However, in the case of the real electron beam, such oscillating terms can contribute to the electron trajectory and velocity since they are not less in values than that of terms responsible for the focusing properties of the undulator magnetic field. Indeed, in Eq. (25), the horizontal focusing effect is described by the term $0.25x_0\widetilde{K}^2(z/a)^2 = 0.25x_0\widetilde{K}^2(\varphi/A)^2$. Some terms in Eq. (45) are also proportional to φ^2. For trajectories with the initial conditions $\widetilde{x}_0 \sim \widetilde{K}$, $\widetilde{y}_0 \sim \widetilde{K}$, $\theta_0 \sim \widetilde{K}$ and $y'_0 \sim \widetilde{K}$, these terms are recognised in the same order as the term $0.25\widetilde{x}_0\widetilde{K}^2(\varphi/A)^2$.

The parameter $A = k_z a = 2\pi a/\lambda_u$ describes the degree of nonuniformity of undulator magnetic field along the horizontal axis, see Eqs. (4)–(6). For a planar undulator with infinitely wide poles, we have $a \rightarrow \infty$ and consequently $A \rightarrow \infty$. In this case, some terms in Eqs. (45), (46) become zero, and these formulas are simplified significantly.

6. Analysis of the obtained results

In the earlier sections, we have derived two sets of formulas which describe electron trajectories in the ideal field of a planar undulator. The first set, given by Eqs. (25)–(28), was derived within the framework of the well-known focusing approximation, and the second set (see Eqs. (45), (46)) was derived by means of the perturbation theory. The electron trajectories in the planar undulator magnetic field can also be simulated numerically by solving Eqs. (2), (3) together with Eqs. (7)–(9) for the three-dimensional undulator field by using the Runge-Kutta algorithm. These electron trajectories once simulated numerically with a small step (which provides high calculation accuracy) can be considered as a reference data for the analysis of the approximate analytical formula precision. In doing so, it is necessary to keep in mind that the numerical solutions of the differential equations of motion also contain some calculation errors. It was demonstrated in the papers [22–24] that numerically computed trajectories, on frequent occasions, differ considerably from the correspondent approximate solutions obtained through the focusing (averaging) approximation. We compare here the numerically simulated electron trajectories with those obtained by using the formulas, derived earlier by methods of perturbation theory, see Eqs. (45), (46) and also with those obtained within the framework of the focusing approximation in accordance with Eqs. (25)–(28).

As an example, we consider the electron trajectories in the undulator at the European XFEL facility (Hamburg, Germany): the reduced electron energy is equal to $\gamma = 35,000$, the undulator period length is $\lambda_u = 40$ mm, the number of undulator periods is $N = 124$, the undulator deflection parameter is $K = 4$, the parameter determining the undulator field nonuniformity along the horizontal axis X (about the pole width), and $a = 50$ mm [28]. The focal lengths for this undulator can be easily found from Eqs. (41), (42): vertical focal length is equal to $f_y = 1231234$ mm and horizontal focal length is equal to $f_x = -77179939$ mm. The horizontal focal length is negative, and the electron beam is defocused by planar undulator along the horizontal direction.

The formulas for electron trajectories, derived in the framework of the focusing approximation (see Eqs. (20), (21), (25), (26)) and by the methods of perturbation theory (Eqs. (45), (46)), include the regular (main) parts of trajectory $x_1(z) = x_0 + \theta_0 z - \left(\widetilde{K}/k_z\right)\sin\varphi$ for the horizontal component and $y_1(z) = y_0 + y'_0 z$ for the vertical component and plus the additional terms, describing the effects of the magnetic field inhomogeneity. Eqs. (25), (26) display it explicitly, while the Eqs. (20), (21), (45), (46) involve these regular parts implicitly, through the trigonometric and hyperbolic functions. Expanding these trigonometric and hyperbolic functions as a power Taylor series in small values $\omega_{x,y}\varphi$, we can easily get the regular parts of trajectory $x_1(z)$ and $y_1(z)$ in explicit form. Clearly, the regular parts $x_1(z)$ and $y_1(z)$ do not describe any focusing properties of planar undulator magnetic field. At the same time, these parts are linear in terms of small parameters $\widetilde{K}$, x_0, y_0, θ_0 and y'_0, while the remaining terms in Eqs. (20), (21), (25), (26), (45), (46) have a cubic degree of smallness and hence are much less than the regular parts.

The regular terms, which are given by Eqs. (14) and (15), are the same for trajectories calculated in the framework of focusing approximation, see Eqs. (25)–(28), and for expressions, derived by the methods of perturbation theory, see Eqs. (45) and (46). For clarity, we consider here the differences $\Delta X(x)$ and $\Delta Y(z)$ between the corresponding solutions and the regular terms:

$$\Delta X(z) = x(z) - x_1(z) \tag{47}$$

$$\Delta Y(z) = y(z) - y_1(z) \tag{48}$$

It is precisely these components that are responsible for the focusing properties of the undulator field.

Figures 2 and **3** show the calculated transversal component (vertical and horizontal correspondently) of electron trajectory and its reduced velocity with the following initial conditions: $x_0 = 0$ mm, $\theta_0 = 0$, $y_0 = +0.1$ mm and $y'_0 = 0$. If so, the regular part of the trajectory (its

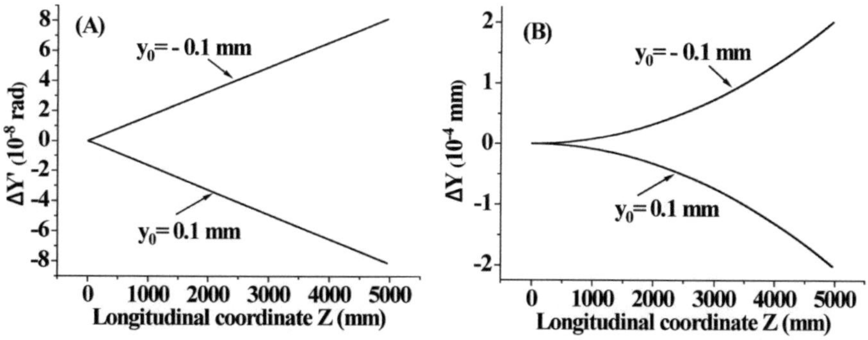

Figure 2. Additions to the electron vertical velocity and vertical coordinate: $x_0 = 0$ mm, $\theta_0 = 0$, $y_0 = \pm 0.1$ mm and $y'_0 = 0$.

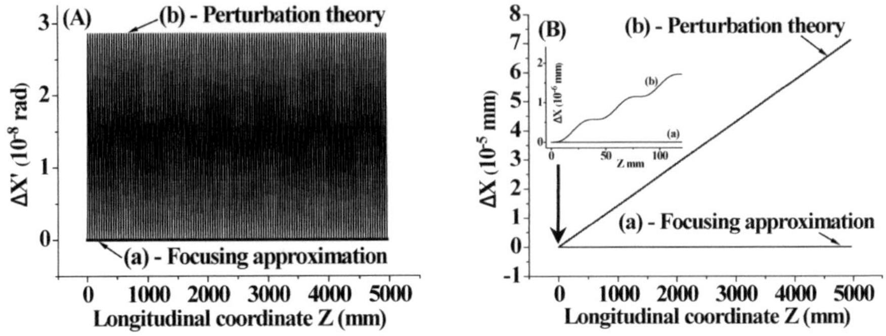

Figure 3. Additions to the electron horizontal velocity and vertical coordinate: $x_0 = 0$ mm, $\theta_0 = 0$, $y_0 = \pm 0.1$ mm and $y'_0 = 0$.

linear approximation) is described by the following expressions: $x_1(z) = -\left(\widetilde{K}/k_z\right)\sin\varphi$ and $y_1(z) = y_0 = \pm 0.1$.

Figure 2 (A) shows the additions $\Delta Y'(z)$ to the linear (regular) parts of the electron-reduced vertical velocities $y'(z)$ for both electrons with $y_0 = \pm 0.1$. Vertical reduced velocity is described by the formula $y'(z) = \Delta Y'(z)$. **Figure 2 (B)** shows the additions $\Delta Y(z)$ to the vertical components of the electron trajectories $y(z)$ calculated at the listed above initial conditions. Vertical component of the electron trajectory is described by the formula $y(z) = y_0 + \Delta Y(z)$. **Figure 2 (A)** and **(B)** both include the addition for the electron trajectory, which was initially shifted downward in the negative direction of Y-axis, $y_0 = -0.1$ mm (the upper curves) and for the electron trajectory, which was initially shifted upward in the positive direction of Y-axis, $y_0 = 0.1$ mm (the lower curves correspondently). It is easy to see that in both cases the electrons in their propagations along the undulator axis deviate monotonically to the undulator median plane $Y = 0$. Indeed, the electron initially located above the median plane (with $y_0 = 0.1$ mm) and moving in parallel to the undulator axis ($y'_0 = 0$) (see **Figure 2 (A, B)**, the lower curves) acquires the negative velocity component, whose absolute value increases linearly with the longitudinal coordinate z growths. Similarly, the electron initially located below the median electron plane (with $y_0 = -0.1$ mm) and moving in parallel to the undulator axis ($y'_0 = 0$) (see **Figure 2 (A, B)**, the upper curves) acquires the positive velocity component, whose value also increases linearly with the longitudinal coordinate z growths. The maximum value of the vertical reduced velocity component is achieved at the final point of the undulator magnetic field and is equal to $\mp 8.11648 \cdot 10^{-8}$ rad. Hence, we can calculate the vertical focal length $f_y \cong (-y_0)/\left(\beta_y(z = \lambda_u N)\right) \cong 1232061$ mm. This coincides with a very small relative error of about 10^{-3} with the analytical value obtained earlier by using Eq. (38).

In fact, all three methods of calculation, namely focusing approximation, perturbation theory and numerical simulation, give just the same result in this case. All correspondent curves in **Figure 2** are merged together. The largest absolute difference between numerically simulated function $\Delta Y'(z)$ by using the Runge-Kutta method and calculated in the focusing approximations is

$6{\cdot}10^{-11}$. In this case, the maximum value of the function $\Delta Y'(z)$ is equal to $8{\cdot}10^{-8}$. Therefore, the relative error is about $8{\cdot}10^{-4}$. The largest absolute difference between the numerically simulated function $\Delta Y'(z)$ and those using formula (46) is equal to 10^{-11}. The relative error in this case is about 10^{-4}. This is better by an order as compared with the case of focusing approximation, although it is not essential in this case. The maximum value of the function $\Delta Y(z)$ is equal to $2{\cdot}10^{-4}$ mm, see **Figure 2 (B)**. The largest absolute difference between numerically simulated function $\Delta Y(z)$ by using the Runge-Kutta method and calculated in the focusing approximations is $2.5{\cdot}10^{-8}$ mm. The relative error in this case is about 10^{-4}. Maximum absolute difference between the values of $\Delta Y(z)$ obtained by means of the numerical simulation and by using Eq. (46) is also equal to $2.5{\cdot}10^{-8}$ mm with the same relative error of about 10^{-4}.

Figure 3 (A) and **(B)** show the horizontal additions to the regular part of the reduced electron velocity $\Delta X'(z)$ and electron trajectory $\Delta X(z)$, calculated at the mentioned earlier conditions $x_0 = 0$ mm, $\theta_0 = 0$, $y_0 = \pm 0.1$ mm and $y'_0 = 0$. The functions $\Delta X'(z)$ and $\Delta X(z)$ are the same for $y_0 = \pm 0.1$ mm since the undulator magnetic field amplitude increases symmetrically with respect to its median plane. The focusing approximation gives zero results for this case: $\Delta X'(z) = 0$ and $\Delta X(z) = 0$. The zero result for $\Delta X'(z)$ and $\Delta X(z)$, given by the focusing approximation, is clear because the equations of motion in the horizontal and vertical planes are independent in the framework of this approximation. Therefore, the shift of the electron in the vertical direction leads to a corresponding variation of the vertical component of its trajectory without changing its horizontal component. At the same time, the numerical simulations by using the Runge-Kutta method, as well as the analytical calculations by using a more exact formula, namely the derivative of Eq. (45), give distinctly nonzero result. In other words, more accurate numerical simulations and analytical calculations carried out with more precise Eq. (45) have clearly demonstrated that electron trajectories in undulators have the cross-coordinate influence effects. This means that changes in the initial electron parameters in the vertical plane lead to changes of the electron trajectory in the horizontal plane, and vice versa. Eqs. (45), (46) demonstrate it explicitly. Physics of the appearance of such nonzero oscillating behaviour of additions (curves (b) in **Figure 3 (A)** and **(B)**) is clear. The undulator deflection parameter K is calculated using the value of the magnetic-field amplitude B_0 on the undulator axis (Z-axis). The transverse electron velocity with zero initial conditions in its motion in the undulator median plane is described by the expression $x'(z) = -\widetilde{K}\cos\varphi$. In the example under consideration, the electron moves at $y_0 = \pm 0.1$ mm (the additional shift in the vertical direction acquired by the electron during its motion in the undulator field is small: $\sim 2{\cdot}10^{-4}$ mm and can be neglected, see **Figure 2 (B)**. The magnetic field in the plane of the electron motion $y_0 = \pm 0.1$ mm is slightly larger than it is in the undulator median plane. As a result, the amplitude of such an electron oscillation must be larger than that if it moves in the undulator median plane. The nonzero addition to the reduced velocity $\Delta X'(z)$ shown in **Figure 3 (A)** describes the increase in the amplitude of the electron-velocity oscillations in the horizontal plane. We also note that in the case under consideration, the additions to the horizontal components of the electron-reduced velocity $\Delta X'(z)$ and coordinate $\Delta X(z)$ (see **Figure 3 (A), (B)**) are of the same order in amplitude as the corresponding additions to the vertical components $\Delta Y'(z)$ и

$\Delta Y(z)$, see **Figure 2 (A)**, **(B)**: $\Delta Y'(z) \sim \Delta X'(z)$, $\Delta Y(z) \sim \Delta X(z)$. The ration of amplitudes of these functions is about 2.5 only.

Functions $\Delta X'(z)$ and $\Delta X(z)$, calculated through Eq. (45) and its derivative, are in excellent agreement with these numerically simulated functions. The function $\Delta X'(z)$, differently calculated (numerically and with the Eq. (45)), has maximum deviation about $6{\cdot}10^{-11}$ in its absolute value. The maximum value of $X'(z)$ function is about $3{\cdot}10^{-8}$ and the relative error is equal to $2{\cdot}10^{-3}$. Similarly, the function $\Delta X(z)$, differently calculated, has maximum deviation about 10^{-9} in its absolute value. The maximum value of $\Delta X(z)$ function is about $7{\cdot}10^{-5}$ and the relative error is equal to $1.4{\cdot}10^{-5}$. This means that these functions are merged together in **Figure 3 (A)** and **(B)**.

Figures 4 and **5** show the calculated transverse components of the electron trajectory and its reduced velocity $\Delta X(z)$, $\Delta Y(z)$, $\Delta X'(z)$ and $\Delta Y'(z)$ with the following initial conditions: $x_0 = 0.1$ mm, $\theta_0 = -5{\cdot}10^{-5}$ rad, $y_0 = 0.1$ mm and $y'_0 = -5{\cdot}10^{-5}$ rad.

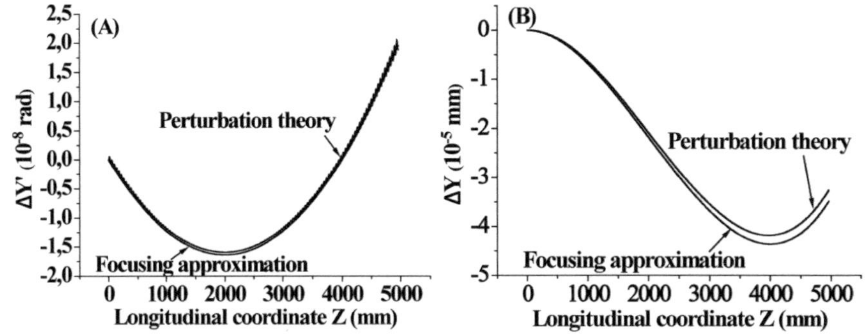

Figure 4. Additions to the electron vertical velocity and vertical coordinate: $x_0 = 0.1$ mm, $\theta_0 = -5{\cdot}10^{-5}0$ rad, $y_0 = 0.1$ mm and $y'_0 = -5{\cdot}10^{-5}$ rad.

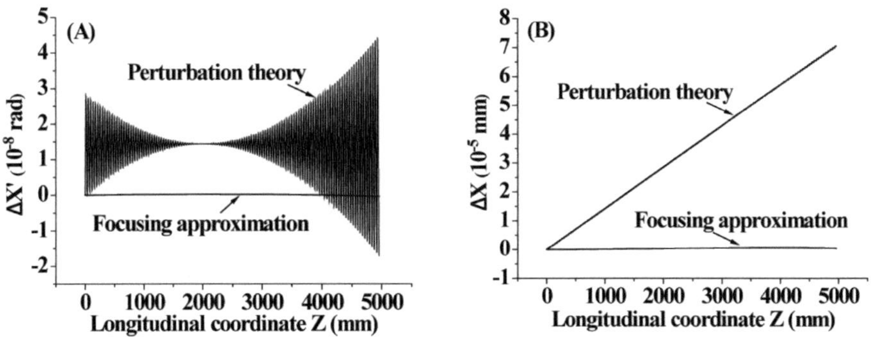

Figure 5. Additions to the electron horizontal velocity and horizontal coordinate: $x_0 = 0.1$ mm, $\theta_0 = -5{\cdot}10^{-5}0$ rad, $y_0 = 0.1$ mm and $y'_0 = -5{\cdot}10^{-5}$ rad.

Figure 4 shows the additions to the linear (main) part of the vertical component of the electron-reduced velocity $\Delta Y'(z)$ and coordinate $\Delta Y(z)$. It can be seen that both curves in **Figure 4 (A)**, as well as in **Figure 4 (B)**, namely computed in the framework of focusing approximation and by means of perturbation theory, are relatively close to each other. In most cases, the difference is not important. For completeness, we check here the precision of Eq. (46). For **Figure 4 (A)**, the largest absolute difference between the numerically simulated function $\Delta Y'(z)$ and those using formula (46) is equal to $3{\cdot}10^{-12}$. In this case, the maximum value of the function $\Delta Y'(z)$ is equal to $2{\cdot}10^{-8}$. The relative error in this case is about $1.5{\cdot}10^{-4}$. Similarly, for **Figure 4 (B)** the largest absolute difference between the numerically simulated function $\Delta Y(z)$ and those using formula (46) is equal to $4{\cdot}10^{-9}$ mm. In this case, the maximum value of the function $\Delta Y(z)$ is equal to $4{\cdot}10^{-5}$ mm. The relative error in this case is about 10^{-4}. This means that these two couples of functions are merged together in **Figure 4 (A)** and **(B)**.

Figure 5 shows the additions to the linear (main) part of the horizontal component of the electron-reduced velocity $\Delta X'(z)$ and coordinate $\Delta X(z)$. For **Figure 5 (A)**, the largest absolute difference between the numerically simulated function $\Delta X'(z)$ and those using formula (45) is equal to $1.5{\cdot}10^{-11}$. In this case, the maximum value of the function $\Delta X'(z)$ is equal to $4{\cdot}10^{-8}$. The relative error in this case is about $4{\cdot}10^{-4}$. Similarly, for **Figure 5 (B)**, the largest absolute difference between the numerically simulated function $\Delta X(z)$ and those using formula (45) is equal to $9{\cdot}10^{-10}$ mm. In this case, the maximum value of the function $\Delta X(z)$ is equal to $7{\cdot}10^{-5}$ mm. The relative error in this case is about 10^{-5}. This means that these two couples of functions are merged together in **Figure 5 (A)** and **(B)**. It is seen that, in this case, the focusing approximation describes the electron trajectory in the horizontal plane completely incorrectly, while formula (45) describes it with very good accuracy.

7. Conclusion

Here, electron beam dynamics in a planar undulator was analysed. Three methods of electron trajectory calculations were considered: smoothing (focusing) approximation, perturbation theory method and numerical simulations by using the Runge-Kutta algorithm. Within the framework of focusing approximation, trajectories were described by rather simple analytical expressions (20–23) which have a clear physical interpretation. However, the more detailed analysis of the electron trajectories in a three-dimensional magnetic field of a planar undulator showed that the focusing approximation does not always give the correct result, and it should be used with caution. Expressions (45, 46) give the correct result and their high accuracy was confirmed by numerical simulations. However, Eqs. (45), (46) are rather cumbersome, and they have no clear physical interpretation. Their cumbersomeness results from the fact that they include all terms of cubic power of smallness.

The examples used in this chapter show that the focusing approximation formulas (21, 23), which describe the electron motion in the vertical plane of ideal undulator magnetic field, have quite admissible accuracy. However, in the general case, formulas (20, 22) are hardly applicable to the description of the behaviour of an electron in the horizontal plane. The use of expression (46) gives a more reliable result. At the same time, the use of analytical expressions (45, 46) has significant advantages. Indeed, for numerical calculation (e.g. by using the Runge-Kutta algorithm) the spatial coordinates and velocity directions of an electron at the undulator end, it is necessary to calculate all its trajectories in the undulator successively, step by step, with a small interval. This requires considerable time. By using the analytical formulas, it is possible to immediately obtain the final result by substituting the ending coordinate of the undulator magnetic field into the analytical expressions. This dramatically reduces the computational time.

Acknowledgements

The author is grateful for the support from the joint German-Russian project EDYN_EMRAD, in agreement with the Ministry of Education and Science of the Russian Federation No. 14.587.21.0001, unique identifier of scientific research RFMEFI58714X0001.

Author details

Nikolay Smolyakov[1,2]*

*Address all correspondence to: smolyakovnv@mail.ru

1 National Research Center "Kurchatov Institute", Moscow, Russia

2 Moscow Physical-Technical Institute (State University), Moscow, Russia

References

[1] Barkov LM, Baryshev VB, Kulipanov GN, Mezentsev NA, Pindyurin VF, Skrinsky AN, Khorev VM. A proposal to install a superconducting wiggler magnet on the storage ring Vepp-3 for generation of the synchrotron radiation. Nuclear Instruments and Methods. 1978;**152**

[2] Walker RP. Electron beam focusing effects and matching conditions in plane periodic magnets. Nuclear Instruments and Methods. 1983;**214**:497-504

[3] Torggler P, Leubner C. Accurate analytic off-axis electron trajectories in realistic two-dimensional transverse wigglers with arbitrary magnetic field variation. Physical Review A. 1989;**39**:1989-1999

[4] Smolyakov NV. Focusing properties of a plane wiggler magnetic field. Nuclear Instruments and Methods in Physics Research. 1991;**A308**:83-85

[5] Smolyakov NV. Focusing of a beam of relativistic particles in the magnetic field of a wiggler. Soviet Physics Technical Physics. 1992;**37**(3):309-313

[6] Elleaume P, Chavanne J. A new powerful flexible linear/helical undulator for soft X-rays. Nuclear Instruments and Methods in Physics Research. 1991;**A304**:719-724

[7] Elleaume P. Insertion devices for the new generation of synchrotron sources: A review. Review of Scientific Instruments. 1992;**63**:321-326

[8] Elleaume P. A new approach to the electron beam dynamics in undulators and wigglers. In: Proceedings of the Third European Particle Acceleration Conference (EPAC'92); 24–28 March 1992; Berlin. Singapore: Editions Frontieres; 1992. p. 661-663. Available from: http://accelconf.web.cern.ch/AccelConf/e92/PDF/EPAC1992_0661.PDF

[9] Barnard JJ. Anharmonic betatron motion in free electron lasers. Nuclear Instruments and Methods in Physics Research. 1990;**A296**:508-515

[10] Scharlemann ET. Wiggle plane focusing in linear wigglers. Journal of Applied Physics. 1985;**58**(6):2154-2161

[11] Dattoli G, Renieri A. Experimental and theoretical aspects of the free-electron laser. In: Stitch ML, Bass M, editors. Laser Handbook. Vol. Volume 4. Amsterdam, Oxford, New York, Tokyo: Horth-Holland; 1985. pp. 1-141

[12] Bizzarri U, Ciocci F, Dattoli G, De Angelis A, Fiorentino E, Gallerano GP, Letardi T, Marino A, Messina G, Renieri A, Sabia E, Vignati A. The free-electron laser: Status and perspectives. Rivista Del Nuovo Cimento. 1987;**10**(5):1-131

[13] Serov AV. Influence of the inhomogeneity of electromagnetic wave and undulator fields on the operation of a free-electron laser. Soviet Journal of Quantum Electronics. 1985;**12**: 338-342

[14] Serov AV. Interaction of a relativistic beam with an inhomogeneous electromagnetic wave. Soviet Journal of Quantum Electronics. 1989;**19**(3):351-355

[15] Smith L. Effects of Wigglers and Undulators on Beam Dynamics. Lawrence Berkeley Laboratory: University of California Internal Report LBL-21391; 1986. 3 p

[16] Nagaoka R, Yoshida K, Tanaka H, Tsumaki K, Hara M. Effect of insertion devices on beam dynamics of the 8 GeV light source storage ring. In: Proceedings of the 1989 IEEE Particle Acceleration Conference (PAC 1989); 20–23 March 1989; Chicago, IL, USA. New York: IEEE; 1989. p. 1361-1363. Available from: http://accelconf.web.cern.ch/AccelConf/p89/PDF/PAC1989_1361.PDF

[17] Karantzoulis E, Nagaoka R. Effects of insertion devices on beam dynamics in the presence of closed orbit distortions. In: Proceedings of the 2nd European Particle Acceleration Conference (EPAC'90); 12–16 June 1990; Nice France. Gif-sur-Yvette Cedex - France: Editions Frontieres; 1990. p. 1414-1416. Available from: http://accelconf.web.cern.ch/AccelConf/e90/PDF/EPAC1990_1414.PDF

[18] Einfeld D, Levichev E, Piminov P. Influence of insertion devices on the ALBA dynamic aperture. In: Proceedings of the 11th European Particle Accelerator Conference (EPAC'08); 23–27 June 2008; Genoa. p. 2279-2281. Available from: http://accelconf.web.cern.ch/AccelConf/e08/papers/wepc117.pdf

[19] Tomin S, Korchuganov V. Insertion devices influence on the beam dynamics at Siberia-2 storage ring. In: Proceedings of the fourth International Particle Accelerator Conference (IPAC'13); 12–17 May 2013; Shanghai. p. 193-195. Available from: http://accelconf.web.cern.ch/AccelConf/IPAC2013/papers/mopea051.pdf

[20] Korchuganov VN, Svechnikov NY, Smolyakov NV, Tomin SI. Special-purpose radiation sources based on the Siberia-2 storage ring. Journal of Surface Investigation: X-ray, Synchrotron and Neutron Techniques. 2010;**4**(6):891-897. DOI: 10.1134/S1027451010060030

[21] Altarelli M, Brinkmann R, Chergui M, et al. The European X-Ray Free Electron Laser Technical Design Report. DESY XFEL Project Group 2006–097: Hamburg; 2006. 630 p

[22] Smolyakov N, Tomin S, Geloni G. Electron trajectories in a three-dimensional undulator magnetic field. In: Proceedings of the fourth International Particle Accelerator Conference (IPAC'13); 12–17 May 2013; Shanghai. p. 2223-2225. Available from: http://accelconf.web.cern.ch/AccelConf/IPAC2013/papers/wepwa044.pdf

[23] Smolyakov N, Tomin S, Geloni G. Analytical and numerical analysis of electron trajectories in a 3-D undulator magnetic field. In: Proceedings of the 35th International Free-Electron Laser Conference (FEL2013); 26–30 August 2013; New York, USA. pp. 406-409. Available from: http://accelconf.web.cern.ch/AccelConf/FEL2013/papers/tupso77.pdf

[24] Smolyakov NV. Theoretical and numerical analysis of electron motion in a three-dimensional Undulator field. Journal of Surface Investigation: X-ray, Synchrotron and Neutron Techniques. 2016;**10**(5):1016-1022. DOI: 10.1134/S1027451016050396

[25] Smolyakov N, Tomin S, Geloni G. Electron motion in a 3-D undulator magnetic field. Journal of Physics: Conference Series. 2013;**425**:032023, 4. DOI: 10.1088/1742-6596/425/3/032023

[26] Steffen K. Fundamentals of acceleration optics. In: Turner S, editor. Synchrotron Radiation and Free Electron Lasers. Proceedings of CAS CERN Accelerator School CERN 90–03; 6–13 April 1989; Chester, United Kingdom. Geneva; 1990. p. 1-23

[27] Steffen KG. High Energy Beam Optics. New York - London – Sidney: Interscience publisher; 1965. 211 p

[28] Tschentscher Th. Layout of the X-Ray Systems at the European XFEL. XFEL.EU TN-11-001. European X-Ray Free Electron Laser Facility GmbH: Albert-Einstein-Ring 19 Hamburg; 2011. 21 p

Section 2

Advanced Accelerator Concepts

Concept and Numerical Simulations of Multi-Beam Linear Accelerators EVT with Depressed Collector for Drive Beam

Vladimir E. Teryaev

Additional information is available at the end of the chapter

http://dx.doi.org/10.5772/intechopen.70956

Abstract

The concept of the electron multi-beam linear accelerator electron voltage transformer (EVT) invokes aspects of a two-beam accelerator. Here, effective transformation of energy of several drive beams to energy of a high voltage accelerated beam takes place. It combines an RF generator and essentially an accelerator within the same vacuum envelope. Acceleration occurs in inductively-tuned cavities. The ratios between values of RF beam currents and their voltages are similar to ratios for an electric transformer that is leaded to dub this device an electron voltage transformer (EVT). Properties and high efficiency of the EVT accelerator proves to be true by the example of numerical simulations of three accelerator configurations generating electron beams with energies 1, 2 and 4.3 MeV, respectively. The submitted results of numerical simulations show a resource for the further increase in efficiency of the EVT accelerator up to 70–75% by using a depressed collector for a spent dive beam.

Keywords: two-beam accelerator, electron gun, drive beam, accelerated beam, RF buncher, accelerating cavity, depressed collector

1. Introduction

Two-beam linear collider can serve as the most known example use of two-beam concept in the field of high energy physics [1]. Derbenev Ya et al. [2] can serve as an example of consideration of the two-beam concept in the range of low energy (1–10 MeV). Specificity of low energy when the relativistic factor is low is that a change of velocity of particles for conservation of the appropriate phase ratio between drive bunches and accelerated bunches, i.e. for keeping of synchronism in process of beam-beam interactions, shall be taken into account. Derbenev Ya et al. assumed that the use of the H_{020} mode in a pill-box cavity where acceleration occurs on an axis of the cavity and the annular drive beam changes its radius during movement [2]. Synchronization is carried out due to change in a focusing magnetic field along the axis of the device and this implies changing of the radius of the annular drive beam. However, this concept has never been developed any further.

Here, a new concept of the electron accelerator briefly named electron voltage transformer (EVT) is considered, where the multi-beam approach is of a crucial importance. The approach to use several beam-lets interacting with the common cavity is well-known and is implemented successfully in multi-beam klystrons (MBK's) [3, 4]. Such design is also inherent to the greatest degree in the concept of EVT. All beams in EVT are generated in a common electron gun. Then, they are bunched in RF cavities as in a klystron. The energy is transferred from the drive bunches to the accelerated bunch in the accelerating structure. Each cavity of the structure is coupled to all beams, but the cavities are not coupled one to another.

Basic difference from the concept quoted above [2] is that axes of all beams are parallel straight lines and that for beams focusing a homogeneous magnetic field of a solenoid is used. The cavities in EVT are inductively detuned, having the resonant frequencies of their working modes being higher than the operating frequency. As a result, the phase angle between the gap voltage of the cavities and the fundamental harmonic of the drive beam current is close to 90°, as against conditions for the accelerated beam where the phase angle between accelerating gap voltage and a harmonic of a current is close to zero. This operating regime allows one to achieve the desired value of the accelerating field. The most important advantage of using inductively tuned cavities is effective longitudinal bunch focusing, which is exerted by the cavity RF fields of the drive bunches. The drive beam-lets in EVT are weakly relativistic. Therefore, synchronous energy transfer from the drive beams to the accelerated beam is achieved by correct choices of distances Δz between the adjacent cavity gaps. A similar condition, of two-beam acceleration in inductive tuned cavity, is considered by Kazakov et al. [5], where, in contrast to EVT, the authors have placed the drive beam and the accelerated beam on the same axis. Dolbilov G has discussed a two-beam accelerator of protons, where the use of inductive tuned cavities for an electron drive beam is considered [6, 7]. The accelerated beams and the drive beams move there in opposite directions in order to maintain the condition of synchronization.

Preliminary results of numerical simulations using the known and reliable code MAGIC [8] are presented for an EVT prototype, which operates in S-band, where acceleration up to 1 MeV was predicted at efficiency of 66% [9]. A project version of an EVT accelerator operating at L-band, in which a 6-A, 110-kV beam is accelerated to 2 MeV with an overall efficiency of 60%, is presented below. The further development of the EVT concept is based on a principle of two-stage acceleration, where an ensemble of the drive beam-lets is consisting of two groups. At the first stage, the first group of drive beam-lets transfers their energy to the second group of drive beam-lets. At the second stage, the second group of drive beam-lets transfers their energy to the accelerated beam. Such configuration allows a transformation ratio and energy of the accelerated beam to be increased considerably. Numerical simulations predict acceleration of an initial 1 A, 60 keV beam up to energy of 4.3 MeV at efficiency of 22%. The use of a novel view on a Depressed Collector (DC) for a spent drive beam, where the DC is considered as a separator of a spectrum of particles, increases efficiency in the first two EVT projects up to 74 and 68%, respectively. Numerical results are shown in Figures and Tables.

A positive feature of the EVT concept is that the accelerator does not require an external high power RF-source, but only a 60–110-kV pulsed source for operation. The duty cycle for the device will be easily adjusted via the pulse duty cycle of the power supply. The absence of an RF-source, utilization of a relatively low-voltage pulsed power supply to drive the machine, and the compact size of the device would ensure a relatively light weight and highly versatile applied accelerators.

2. EVT configuration

The schematic representation of an EVT accelerator is shown in **Figure 1**, where the accelerating structure is pop up.

An electron gun, a magnetic focusing system and a RF buncher will be provided below in the description of specific projects of EVTs. **Figure 1** (right) provides an example of the accelerating cavity. The operating mode is TM_{020}. An accelerated beam is located in the centre, and a drive beam is located on the periphery, at a peak value of an electric field. The drive beam consists of six beam-lets in the case shown. An additional passive ring-shaped bunching cavity with the fundamental TM_{010} mode is located in a long drift tube. Synchronous energy transfer from the drive beams to the accelerated beam can be achieved through correct choices of distances Δz_1 and Δz_2 between gaps of adjacent accelerating cavities.

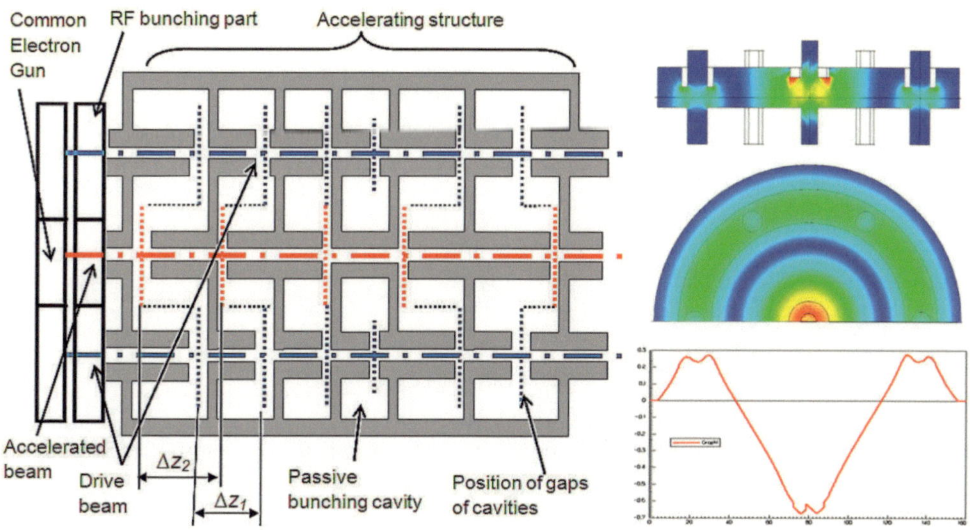

Figure 1. Schematic representation of an EVT (left); shape of the accelerating cavity and electric field E_z along x-axis (right).

3. Scientific and technical justification of the EVT concept

It is known that a field $\mathbf{E}(\mathbf{r},t)$ in a volume Ω of a cavity generated by a beam with a current density $\mathbf{J}(\mathbf{r},t)$ can be presented as eigenmode expansion: $\mathbf{E}(\mathbf{r},t) = \sum_\lambda e_\lambda(t)\cdot\mathbf{E}_\lambda(\mathbf{r})$, where eigen vectors $\mathbf{E}_\lambda(\mathbf{r})$ are functions of coordinate $\mathbf{r}$ that are normalized by energy of the cavity as follows:

$$W_\lambda = \frac{\varepsilon_0}{2}\int_\Omega |\mathbf{E}_\lambda(\mathbf{r})|^2 d\Omega. \tag{1}$$

A time-dependent multiplier $e_\lambda(t)$ is the solution of the equation of oscillator with the second member, Eq. (2):

$$\frac{d^2e_\lambda}{dt^2} + \frac{\omega_\lambda}{Q_\lambda}\frac{de_\lambda}{dt} + \omega_\lambda^2 e_\lambda = -\frac{1}{2W_\lambda}\frac{d}{dt}\left(\int_\Omega \mathbf{j}\cdot\mathbf{E}_\lambda d\Omega\right) \tag{2}$$

Let's express Eq. (2) in a complex form and consider one of the eigenmodes only, having frequency ω_0, replacing: $e_\lambda(t)\to\dot{e}_0\cdot\exp(i\omega t)$, $\mathbf{J}(\mathbf{r},t)\to\mathbf{J}(\mathbf{r})\exp(i\omega t)$. It gives

$$\dot{e}_0 = \frac{i\cdot\omega}{\omega^2-\omega_0^2 - i\cdot\omega\omega_0/Q_0}\cdot\frac{1}{2W_0}\int_\Omega \dot{\mathbf{J}}(\mathbf{r})\mathbf{E}_0(\mathbf{r})d\Omega \tag{3}$$

Other terms of expansion generated by the beam current will have small values at high quality Q_0 of the cavity and at $\omega\approx\omega_0$, and they will not be taken into account. Eqs. (2) and (3) are the most general expressions for determination of fields in cavities [10]. As it takes place $\omega\approx\omega_0$, the equation Eq. (3) can be rewritten as:

$$\dot{e}_0 = \exp(i\varphi)\frac{Q_0}{\sqrt{1+\xi_0^2}}\cdot\frac{1}{2\omega W_0}\int_\Omega \dot{\mathbf{j}}\cdot\mathbf{E}_0 d\Omega \tag{4}$$

where ξ_0 is a relative detuning:

$$\begin{cases} \xi_0 = Q_0\dfrac{\omega^2-\omega_0^2}{\omega\omega_0}\approx\dfrac{2Q_0\Delta\omega_0}{\omega} \\ \Delta\omega_0 = \omega-\omega_0 \\ \varphi = \tan^{-1}(-\xi_0) \end{cases} \tag{5}$$

Practically, the cavity is described via integral characteristics, such as geometric impedance of drive gap (index 1) and accelerated gap (index 2):

$$\rho_1 = (R/Q)_1,\ \rho_2 = (R/Q)_2; \tag{6}$$

circuit current $\dot{I}_1$ of the drive gap and circuit current $\dot{I}_2$ of the accelerating gap equal to

$$\dot{I}_1 = \exp(i\varphi_1) T_{G1} \cdot 2F_{B1} I_{01}, \qquad \dot{I}_2 = \exp(i\varphi_2) T_{G2} \cdot 2F_{B2} I_{02}; \tag{7}$$

RF beam currents:

$$I_{\text{RF1}} = 2F_{B1} I_{01}, \qquad I_{\text{RF2}} = 2F_{B2} I_{02}; \tag{8}$$

voltage $\dot{V}_1$ of the drive gap along axis z_1 and voltage $\dot{V}_2$ of the accelerating gap along axis z_2:

$$\dot{V}_1 = \dot{e}_0 V_1 = \dot{e}_0 \int_{z_1} E_0(z) dz, \quad \dot{V}_2 = \dot{e}_0 V_2 = \dot{e}_0 \int_{z_2} E_0(z) dz \tag{9}$$

where: φ_1 and φ_2 are phase angles; T_{G1}, T_{G2} are transit time factors of the drive and accelerating gaps, respectively; F_{B1}, F_{B2} are form-factors of bunches; I_{01}, I_{02} are average drive beam currents of the drive and accelerated beams.

For relativistic beams, when it is possible to neglect relative moving of particles inside a bunch, and taking into account that the integral on the second member of Eq. (2) is not equal to zero in gap 1 and gap 2 only where beam currents exist, the Eq. (2) is rewritten as follows:

$$\dot{e}_0 = \exp(i\varphi) \frac{Q_0}{\sqrt{1+\xi_0^2}} \cdot \frac{1}{2\omega W_0} \left(\dot{I}_1 V_1 + \dot{I}_2 V_2 \right). \tag{10}$$

Let us add in relationships under consideration

$$K_V = V_2 / V_1 = \sqrt{\rho_2 / \rho_1}, \tag{11}$$

following from: $2\omega W_0 = V_1^2/\rho_1 = V_2^2/\rho_2$.

The description of the equivalent circuit for a case of interaction of relativistic beams with the cavity is

$$\dot{V}_1 = \exp(i\varphi) \cdot \frac{Q_0 \cdot \rho_1}{\sqrt{1+\xi_0^2}} \cdot \left(\dot{I}_1 + K_V \dot{I}_2 \right) - Z_C \cdot \dot{I} \tag{12}$$

$$\dot{V}_2 = -K_V \dot{V}_1 \tag{13}$$

where Z_C is impedance of the drive gap, φ is a phase angle between gap voltage $\dot{V}_1$ and equivalent circuit current $\dot{I}$. Eq. (13) shows properties of the chosen operating eigenmode of a cavity when phases of fields in accelerating and drive gaps are shifted by angle π.

At interaction of a cavity with low relativistic beam it is necessary to consider also beam-loading effect or a beam admittance $Y_B = G_B + i \cdot B_B$. Therefore, Eq. (12) is revised as

$$\dot{V}_1 = \frac{1}{1/Z_C + Y_B} \cdot \left(\dot{I}_1 + K_V \dot{I}_2 \right) \tag{14}$$

Let's set out well-known analytical formulas for a rectangular bunch with angular length ψ_B and for the flat cavity gap having a transit time angle ψ_C:

$$T_C = \frac{\sin(\psi_C/2)}{\psi_C/2}; F_B = \frac{\sin(\psi_B/2)}{\psi_B/2}; \tag{15}$$

$$G_B = \frac{I_0}{V_B} \cdot \frac{1}{2} T_G^2 \left(1 - \frac{\psi_C}{2} \text{ctan} \frac{\psi_C}{2}\right); B_B = \frac{I_0}{V_B} \cdot \frac{\sin\psi_C - (1 + \cos\psi_C)\psi_C/2}{\psi_C^2}, \tag{16}$$

where I_0 is an average beam current, V_B is an average beam voltage in the corresponding gap (drive or accelerating), T_G is the corresponding transit time factor and F_B is a form-factor of the corresponding beam.

3.1. Conditions for preset drive bunching maintenance

As it is known, a bunching cavity has inductive detuning, i.e. $\omega < \omega_0$, at which the phase angle between a gap voltage and an equivalent current of a cavity circuit becomes close to $\pi/2$. **Figure 2** shows mutual position of bunches, currents and gap voltages on a phase plane in an accelerating cavity. The phase position of a head $\dot{\boldsymbol{H}}_{head}$ and tail $\dot{\boldsymbol{T}}_{tail}$ of a drive bunch is shown here, too. Let's notice, that complex RF beam current $\dot{\boldsymbol{I}}_{RF1}$ is displaced at angle π relative to the circuit current $\dot{\boldsymbol{I}}_1$.

3.2. An impedance of the drive gap loaded with the accelerated beam

As it is evidently from **Figure 2**, suitable conditions for an accelerating cavity shall be as follows: $\varphi \to \pi/2$, $\exp(i\varphi) \approx \boldsymbol{i}$. It is the chosen phase of drive current: $\varphi_1 = 0$. The phase of the accelerated beam current equal to $\varphi_2 \approx -\pi/2$ provides a bunch being in the electric field close

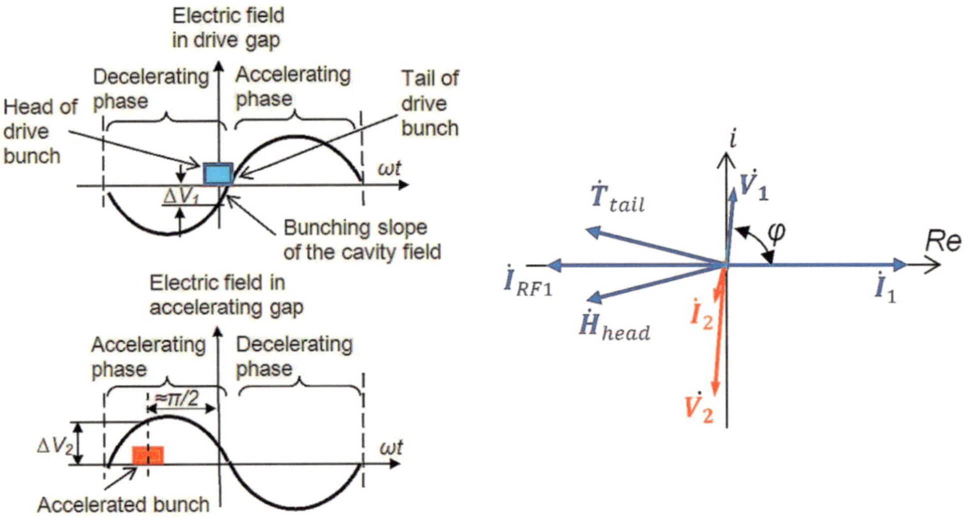

Figure 2. Electric field in cavity gaps (left) and position of complex values on the phase plane (right), providing phase focusing of a drive bunch and acceleration of the beam.

to the peak value, i.e. $\dot{\boldsymbol{I}}_1 = I_1$ and $\dot{\boldsymbol{I}}_2 = -i\cdot I_2$. Thus, using Eqs. (12) and (13), the required impedance of the loaded drive gap is

$$Z_1 = \frac{\dot{\boldsymbol{V}}_1}{\dot{\boldsymbol{I}}_1} = \frac{Q_0\cdot\rho_1}{\sqrt{1+\xi_0^2}}\left(\frac{K_V}{K_{Tr}} + i\right) \tag{17}$$

where $K_{Tr} = I_1/I_2$ is a transformation ratio. An impedance of the accelerating gap is similarly found:

$$Z_2 = \frac{\dot{\boldsymbol{V}}_2}{\dot{\boldsymbol{I}}_2} = \frac{Q_0\cdot\rho_1}{\sqrt{1+\xi_0^2}}\left(K_V K_{Tr} - iK_V^2\right) \tag{18}$$

On the other hand for the equivalent circuit of the cavity the impedance Z_1 is given as:

$$Z_1 = r + i\cdot x = \frac{Q_{load}\cdot\rho_1}{1+\xi_{load}^2}\left(1 - i\cdot\xi_{load}\right) \tag{19}$$

where Q_{load} and

$$\xi_{load} \approx 2Q_{load}\Delta\omega_{load}/\omega \tag{20}$$

are parameters of the equivalent loaded cavity circuit. Equating the real and imaginary parts of Eqs. (18) and (19), the loading quality Q_{load} and relative detuning ξ_{load} can be represented as follows:

$$\begin{cases} Q_{load} = \dfrac{Q_0}{\sqrt{1+\xi_0^2}}\dfrac{K_{Tr}}{K_V}\left(1 + \dfrac{K_V^2}{K_{Tr}^2}\right) = \dfrac{Q_0}{\sqrt{1+\xi_0^2}}\cdot\dfrac{K_{Tr}}{K_V} \\ \xi_{load} = -\dfrac{K_{Tr}}{K_V} \end{cases} \tag{21}$$

Considering Eq. (21) together with Eq. (20) leads to:

$$\frac{\Delta\omega_{load}}{\omega} = -\frac{\sqrt{1+\xi_0^2}}{2Q_0}\cdot\frac{1}{1+K_V^2/K_{Tr}^2} \approx -\frac{|\Delta\omega_0|}{\omega} \tag{22}$$

where a high value of detuning $\Delta\omega_0$ is suggested, which results in $\sqrt{1+\xi_0^2} \approx |\xi_0|$. In addition, it is assumed that the transformation ratio is large enough, i.e., $K_V^2/K_{Tr}^2 \ll 1$. For example, $K_V \approx 2$, $K_{Tr} \approx 20$, $\Delta\omega_0/\omega \approx 10^{-3}$, $Q_0 \approx 8000$, $\xi_0 \approx 16$.

Eq. (22) shows that loading of the accelerated beam practically does not change the imaginary part of the impedance of the cavity, and, thus, it is possible to model loading as a certain equivalent external load quality. This feature is used below in the description of numerical model, see Section 4.

3.3. Voltage transformation ratio

Let's find the voltage ΔV_{B1} lost by the drive beam and the voltage ΔV_{B2} added by the accelerated beam, which are connected to a voltage ΔV_1 of the drive gap and a voltage ΔV_2 of an accelerating gap as assigned at the corresponding moment of time (see **Figure 2**):

$$\Delta V_{B1} \approx T_{G1} \Delta V_1, \quad \Delta V_{B2} \approx T_{G2} \Delta V_2 \tag{23}$$

where transit-time factors T_{G1}, T_{G2} are determined the same as in Eq. (7). Let us find voltages ΔV_1 and ΔV_2 using Eqs. (12) and (13) as follows:

$$\begin{cases} \Delta V_1 = \mathcal{R}e(Z_1) \cdot I_1 = \dfrac{Q_0 \sqrt{\rho_1 \rho_2}}{\sqrt{1+\xi_0^2}} \cdot I_2 \\ \Delta V_2 = \mathcal{R}e(Z_2) \cdot I_2 = \dfrac{Q_0 \sqrt{\rho_1 \rho_2}}{\sqrt{1+\xi_0^2}} \cdot I_1 \end{cases} \tag{24}$$

where the transformation ratio K_{Tr} is:

$$K_{Tr} = \frac{I_1}{I_2} = \frac{\Delta V_2}{\Delta V_1} \tag{25}$$

In view of Eqs. (7), (8), and (23), the transformation ratio can be written through the RF beam currents as:

$$K_{\mathrm{Tr}} = \frac{I_{\mathrm{RF1}}}{I_{\mathrm{RF2}}} \approx \frac{\Delta V_{B2}}{\Delta V_{B1}}, \tag{26}$$

or, at a high value of transit time factors $T_G \to 1$, Eq. (26) can be written as

$$K_{\mathrm{Tr}} = \frac{I_{01}}{I_{02}}, \tag{27}$$

where I_0 is the average current of the corresponding beam. It reflects the law of conservation of energy for the idealized conditions applied in the beginning of section 3.3 and corresponding to small resistive loss in a cavity. In view of losses it is resulted:

$$K_{Tr} = \frac{I_1}{I_2} = \frac{\Delta V_2}{\Delta V_1} \cdot \frac{1}{\eta_B} \tag{28}$$

where η_B is efficiency of power transfer, defined in the Section 3.4.

3.4. Efficiency of power transfer between beams and other parameters of the EVT

From conservation of energy, the efficiency of power transfer from one beam to another can be estimated as:

$$\eta_B = \frac{\Delta V_2 I_2}{\Delta V_2 I_2 + V_2^2/2R_{s2}} \tag{29}$$

where $R_{s2} = Q_0 \cdot \rho_2$ is the shunt impedance of the accelerating gap. For achieving a high efficiency of energy transfer, it is important to have a large enough current I_2 of the accelerated beam, i.e. $I_2 \gg V_2/2R_{s2}$.

Based on the above consideration and **Figure 2**, it can see that in order to reach the effective bunching of drive beam, with an angle between the current $\dot{\boldsymbol{I}}_1$ and the voltage $\dot{\boldsymbol{V}}_1$ close to $\pi/2$, two conditions shall be satisfied. The first is to have a large enough value of the cavity detuning $\Delta\omega_0$; the second is $I_1/K_V \gg I_2$. The angle between $\dot{\boldsymbol{I}}_1$ and $\dot{\boldsymbol{I}}_2$ should be close to $\pi/2$ for effective acceleration since the accelerated bunch should pass through an accelerating gap at the time when the voltage on the gap is near its maximum. Thus, conditions

$$I_1/K_V \gg I_2 \gg V_2/2R_s \tag{30}$$

are the necessary factors of effective performance of the EVT, which are readily achieved when multiple drive beam-lets are used. For example, $I_1 = 60\,A$, $K_V = 2$, $I_2 = 3\,A$, $\Delta V_2 = 80\,kV$, $V_2 = 100\,kV$, $Q_0 \approx 8000$, $\rho_2 = 60\,ohm$, $\eta_B = 0.95$.

We define the efficiency η of the accelerator as a ratio of the power of the accelerated beam at the EVT output to the total beam power in the gun, namely:

$$\eta = \frac{P_{Accel}}{I_{Gun} V_{Gun}} \tag{31}$$

where I_{Gun} is the total current of the gun, and V_{Gun} is the gun voltage. The high EVT efficiency is confirmed by the results of numerical simulations, as shown in **Table 1**.

We define a voltage gain factor to be a ratio of the voltage of the accelerated beam at the EVT output (the final beam energy) to the gun voltage. It is:

$$K_{EVT} = V_{Accel}/V_{Gun} \approx 1 + K_{Tr} \cdot \eta \tag{32}$$

where η and K_{EVT} are shown in the **Table 1**, as a result of numerical simulations.

3.5. Condition of synchronization and phasing of bunches

The drive beam-lets in the EVT are essentially low relativistic. Therefore, a synchronous energy transfer from the drive beams to the accelerated beam is achieved through correct choices of distances Δz between the adjacent accelerating cavities gaps. In case of equality of distances between gaps shown in the **Figure 1**, a condition of synchronism is as follows:

$$\Delta z = \Delta z_1 = \Delta z_2 = \frac{2\pi c}{\omega} \cdot \frac{\beta_1 \beta_2}{\beta_1 - \beta_2} \tag{33}$$

where $\beta_1 c$ and $\beta_2 c$ are velocities of the drive and accelerated bunches in space Δz between cavities gaps. Phasing capabilities considerably extend if distances Δz_1 and Δz_2 between a gaps of adjacent cavities are not equal to each other. More general Eq. (34) is corresponding to this case:

$$\frac{\Delta z_1}{\beta_1} - \frac{\Delta z_2}{\beta_2} = N \cdot \lambda; \quad N = 0; 1; 2; \ldots; \quad \lambda = 2\pi c/\omega \tag{34}$$

It is supposed to use RF bunchers of klystron type for generation of drive bunches. Further bunches should be decelerated in cavities, gradually giving the energy to the accelerated bunches. There is a reasonable question for how long the drive bunches can keep their form

Multi-beam linear accelerator	**S-band**	**L-band**	
Operating frequency, *F*	2856	1300	MHz
Voltage of common electron gun	60	110	kV
Output energy of accelerated beam	1.1	2	MeV
Total gun current of drive beam	72	180	A
Gun current of accelerated beam	3	6	A
Focusing magnetic field in solenoid	800	1000	G
Energy spread of accelerated beam	±10	±10	%
Micro-perveance of one drive beam-let	0.816	0.82	$\mu A/V^{3/2}$
Transmission of drive beam	0.99	0.99	
Transmission of accelerated beam	0.90	0.87	
Transformation ratio	24	30	
Voltage gain factor	≈18	≈18	
Typical R/Q factor of accelerating gap	58	54	Ω
Typical cavity detuning, $\Delta F = F_0 - F$	8 ÷ 12	7	MHz
Typical K_V factor of gaps of cavity (Eq. 11)	4.5	5	
Efficiency of transfer of power (Eq. 29)	94	97	%
Number of beam-lets of drive beam	6	6	
Number of cavities of RF bunchers	4	4	
Number of accelerating cavities	14	12	
Number of passive bunching cavities	1	7	
Length of RF part of accelerator	0.79	2.4	m
Efficiency of accelerator EVT	≈66	≈60	%
Efficiency of EVT with DC for drive beam	72	66	%

Table 1. The simulated parameters of the S-band and L-band linear accelerators EVT.

and phase length. Results of 2D simulations show long enough conservation of the generated bunch (see below projects of EVT). The similar research confirming these results can be found in [6, 7].

4. Representation of 3D objects via equivalent 2D objects

3D codes have been used for simulation of elements of the EVT (cavities, magnetic system), but for the analysis and optimization of dynamics of ensemble of the interacting beams moving in numerous drift tubes the direct and slow 3D analysis is too intricate problem. Difficulty is aggravated also with necessity of carrying out numerous optimizations.

The drive beam-lets are located on a circle, see **Figure 1** (right), and it can be quite replaced with single equivalent 2D annular beam. It is possible to choose the corresponding width of the ring filled with a beam, providing close values of plasma wave-lengths of 2D and 3D models of the beam. In the same way it is replaced real 3D cavities with equivalent 2D cavities. There are chose parameters (gap, *R/Q*) of interaction of a 2D beam with equivalent 2D cavities similar to interaction of a 3D beam with 3D cavities. Certainly, such equivalent model does not describe exact dynamics of the drive beam, but reflects quite well the properties of the EVT accelerator and can serve as initial approach. The results of parameter optimization for the EVT accelerators using the 2D code MAGIC [8], are provided in Sections 5–7.

It is obvious that, on the one hand, implementation of a pure annular concept of the drive beam is embarrassing technically, and, on the other hand, the annular model is rather rough to describe dynamics of a set of cylindrical drive beam-lets. This simulation is provided here to illustrate a method of acceleration used in the EVT as it correctly represents interaction of two electron beams without any additional assumptions.

The following, more adequate model is appropriate for description of a 3D problem, which can be named **2 × 2D** model. This model is based on thesis about an impedance of the loaded drive gap considered in Section 3.3 and on the following circumstances:

1. Each of beam-lets and electric RF fields of cavities near to a beam-let can be considered as axially symmetric relative to a local axis of the given beam-let;
2. Parameters of drive beam-lets are identical;
3. Accelerating cavities are weakly loaded, and, therefore, RF fields of cavities are rather close to a field of an operating RF eigenmode. I.e. the loaded quality of a cavity remains high enough. In other words, the RF power circulating in a cavity $P_{RF} = \frac{V_1^2}{2R/Q_1}$ is much higher than power $\Delta P_2 = P_{output2} - P_{input2}$ transferred to the accelerated beam, for example, $Q_{load} = P_{RF}/\Delta P_2 \geq 100$.

According to these points, it is considered a 3D problem as two connected 2D problems, see **Figure 3**. The drive beam goes through the first equivalent 2D cavity, having the *R/Q* of drive

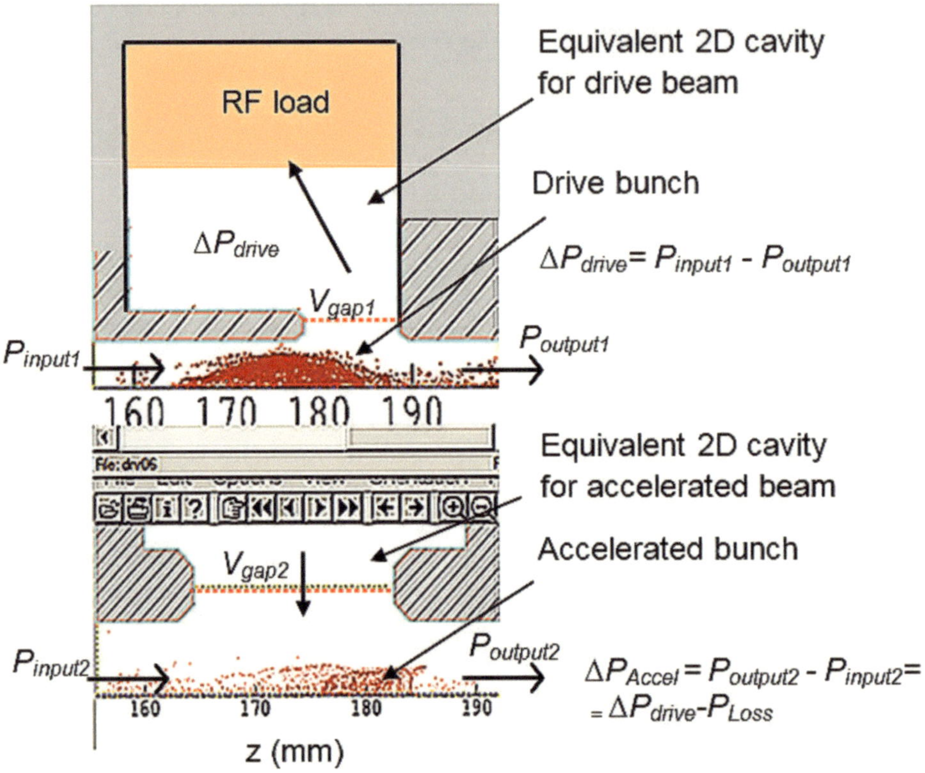

Figure 3. Two 2D cavities, the parameters of which are related via the shown relationships, model processes in a 3D cavity.

gap equal to *R/Q* of corresponding 3D cavity. This cavity has an RF load in which RF power $P_{load} = \Delta P_{drive}$ is absorbed.

The accelerated beam goes through the second cavity having a gap voltage V_{gap2} and a phase angle φ_{gap2}, which are calculated based on a voltage V_{gap1} on the gap of the cavity 1 for the drive beam and where power of the accelerated beam increases at the value $\Delta P_2 = P_{output2} - P_{input2}$. Positions of gaps of these two equivalent cavities correspond to the position of gaps of the 3D cavity. Combination of these two processes will consist in maintenance of equations:

$$\begin{cases} \Delta P_2 = \Delta P_{accel} = 6 \cdot \Delta P_{drive} - P_{loss} \\ V_{gap2} = K_V V_{gap1} \\ \varphi_{gap2} = \varphi_{gap1} + \pi \end{cases} \tag{35}$$

where $N = 6$ is the number of drive beam-lets, P_{loss} is resistive loss in the 3D cavity. Eq. (35) is carried out by choosing the corresponding conductivity of an external RF load of the first (drive) cavity; the value $K_V = \left(\frac{V_{gap2}}{V_{gap1}}\right)_{3D}$ is defined (see Eq. 11) upon the results of the 3D cavity eigenmode analysis. Numerical simulations using the code MAGIC show fast convergence of the Eq. (35) process. 2–3 iterations are practically enough.

5. Design of S-band 1 MeV linear accelerator EVT

The design of a linear accelerator EVT operating in S-band (F = 2856 MHz) and generating 1 MeV electron beam is presented in this section. One accelerated beam and 6 drive beam-lets are generated by a common electron gun fed from a 60 kV power supply. The magnetic lenses around the gun and two matching lenses ensure formation of beams without ripples in the same way, as it is described in Ref. [3]. Bunches are formed in the bunching section energized by two coherent microwaves via RF couplers. The generated drive beam-lets interact with the accelerated beam in the cavities of the energy transfer section (accelerating structure), with the cavities working at the TM_{020} mode. For a layout of the S-band EVT, please see **Figure 4**.

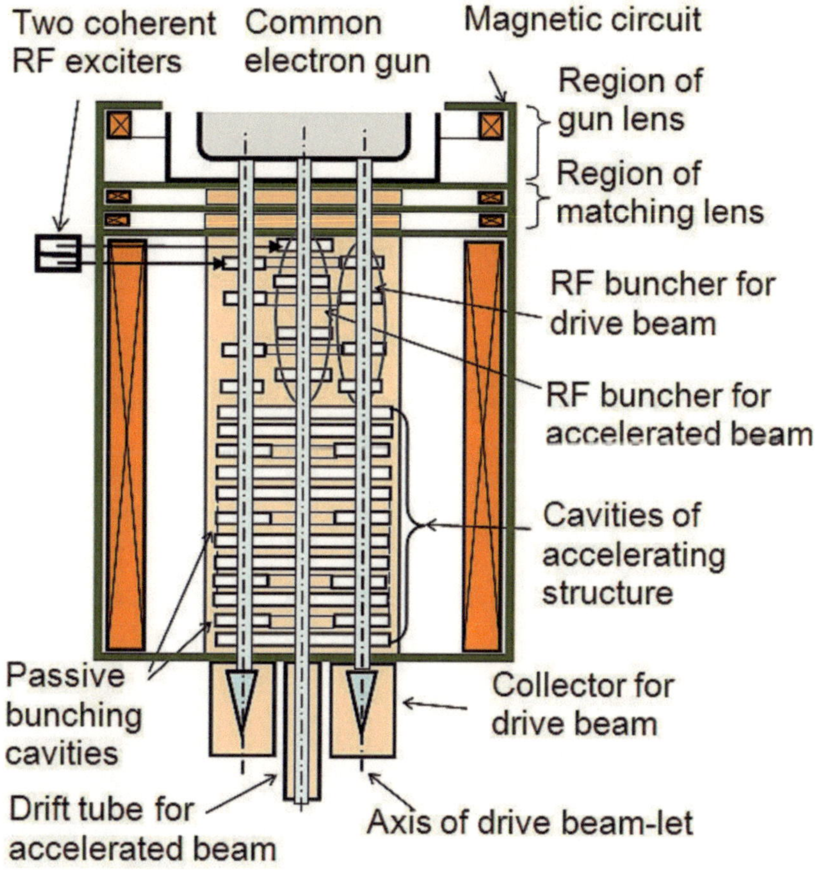

Figure 4. Lay-out of the accelerator EVT.

The results of the initial numerical analysis using code MAGIC are given below in **Table 1**. Two models described above in Section 4 have been applied for the analysis. The models have shown close results. Beam dynamics of the annular model is more evident and results in **Figure 5**. **Figure 6** shows energy distribution of particles of beams in the 1st accelerating cavity ($z > 314$ mm) and at the output of the accelerating structure ($z > 787$ mm). Energy of 1.1 MeV, averaged on the size of a bunch, is achieved. You can see disruption of the spent drive bunches at the accelerator output. The achieved 66% efficiency is twice the best efficiency accessible for the facility consisting of a high power klystron and a high efficient linear accelerator [11].

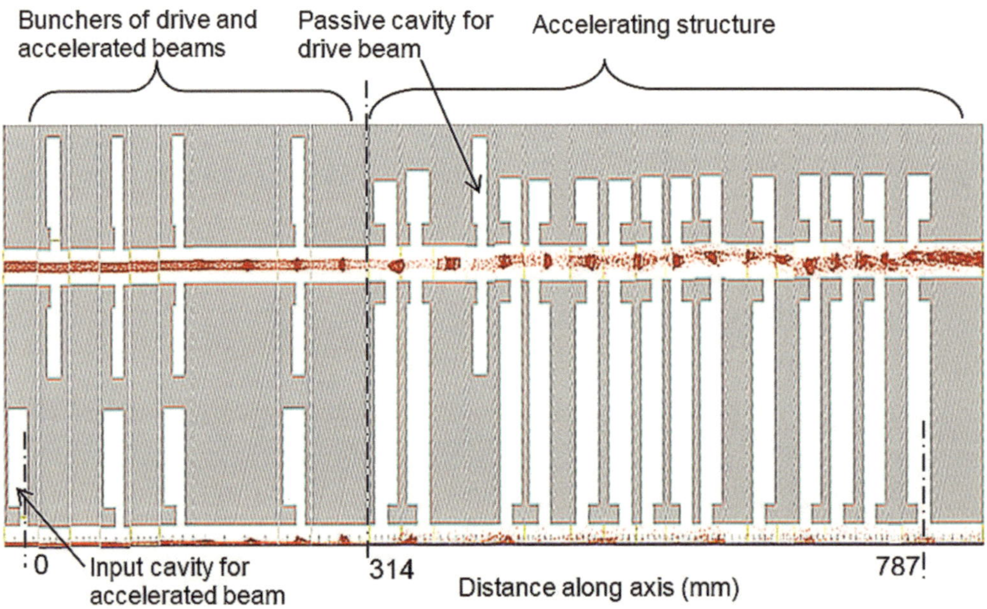

Figure 5. The geometry of a compound RF buncher, accelerating structure and beams dynamics are shown for the S-band EVT.

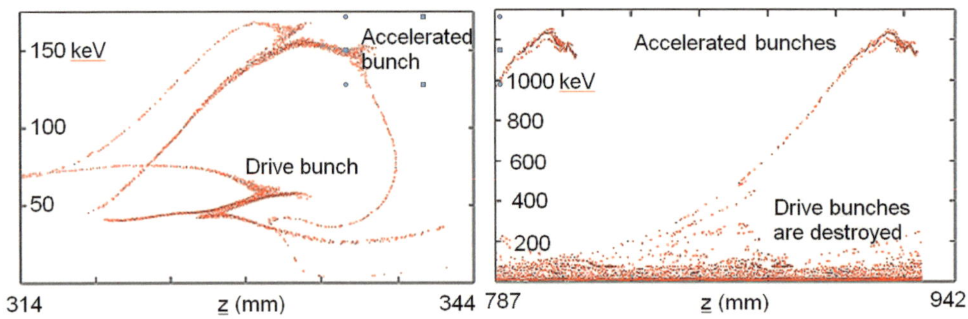

Figure 6. Particle energy distribution in the 1st accelerating cavity (left) and at the output of the accelerating structure (right).

5.1. Features of the multi-beam electron gun

The radius of a ring on which drive beam-lets are located is defined by a frequency and configuration of the cavities (see **Figure 1** (right)). Six cathodes with a current of 12 A are freely located on this radius at the gun voltage 60 kV, **Figure 7**. Clearly, a drive beam-let perveance should not be too high to not interfere with phase stability and synchronization of bunches.

Distance between the drive beam-let cathodes is big enough in comparison with distance between the cathode and anode therefore transverse electrostatic fields on an axis of drive beam-lets are insignificant. Thus the electron optics of each of beam-lets can be quite considered as axial symmetric. **Figure 8** shows the 2D geometry and an example of simulation of a drive beam-let. The double lens around the gun allows obtain the desired profile of the magnetic field to focus the beam-lets and minimize their ripples.

Cells of magnetic lenses are formed by flat iron pole-pieces and system of apertures (see **Figure 4**). Properties of cells for drive beam-lets are identical. The sizes of apertures and distances between

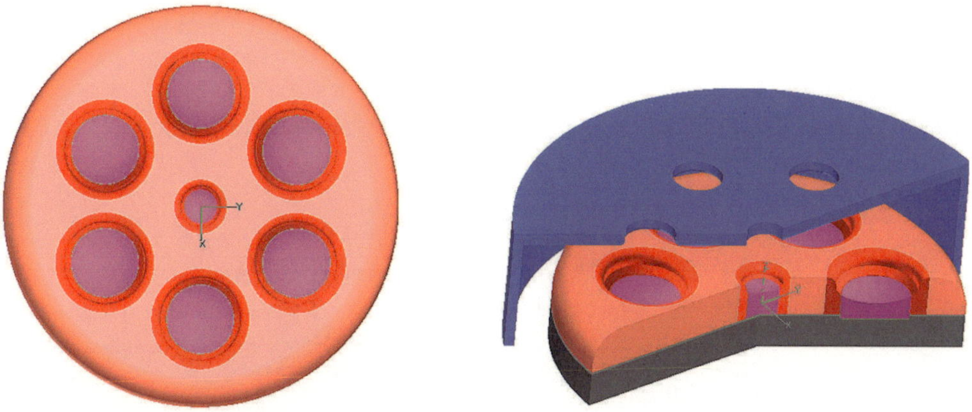

Figure 7. The multi-beam cathode of the electron gun. The peripheral cathodes generate the drive beam-lets, and the central cathode generates the accelerated beam.

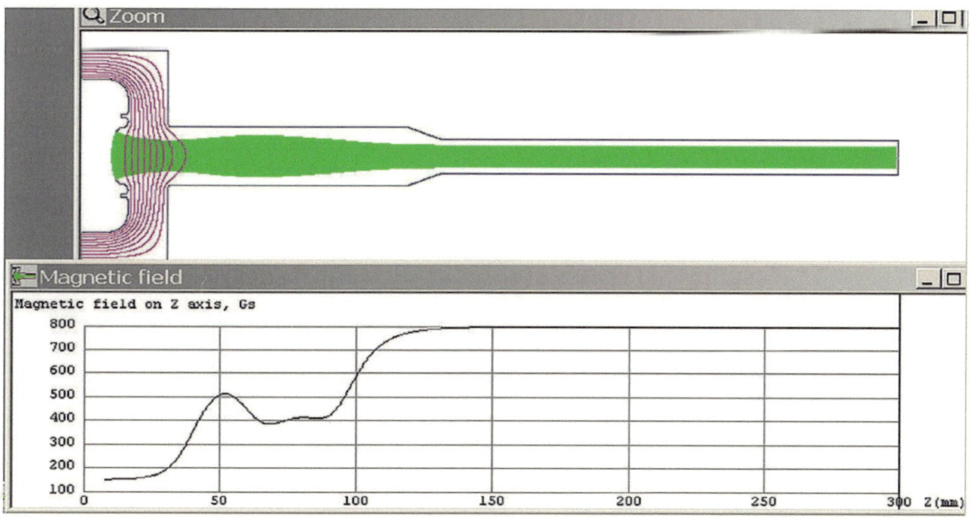

Figure 8. The geometry and an example of 2D simulation of the drive beam-let gun.

them are such that there are no appreciable transverse fields generated on an axis of beam-lets. Thus, the magnetic field quite possesses axial symmetry near the beam-let axis.

6. Specification of L-band 2 MeV linear accelerator EVT

The S-band EVT project has shown achievement of a high efficiency, but has simultaneously shown limited capabilities for an acceleration rate and for a factor of transformation as decay of bunches after $N_{cav} = 15$ cavities has been observed. Numerical results show difficulty of bunch confinement at a voltage of a drive gap $V_{gap1} > \frac{1}{2}V_{gun}$. It results in the following limiting estimated condition:

$$V_{\mathrm{Accel}} < \frac{1}{2} K_V N_{\mathrm{cav}} V_{\mathrm{gun}}, \tag{36}$$

where K_V is from Eq. (11). This estimation is close to observable result $V_{Accel} = 1.1\,MeV$ achieved in the S-band EVT.

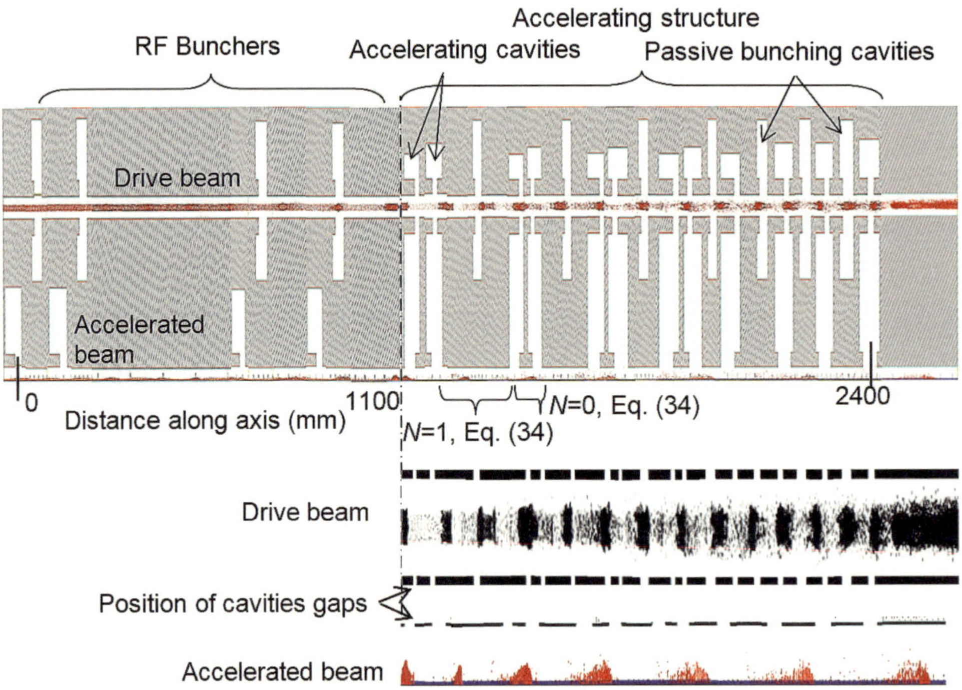

Figure 9. Geometry of the RF structure and the result of numerical simulations are shown for L-band EVT. For more details on the dynamics of the beams, please see below.

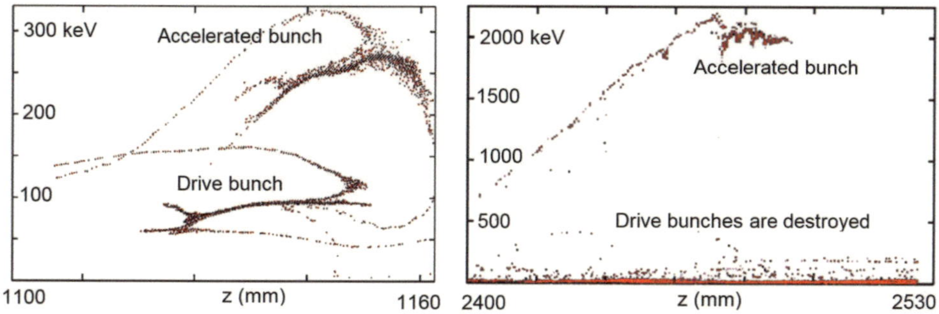

Figure 10. Particle energy distribution at the output of the 1st accelerating cavity (left) and at the output of the accelerator section (right).

Thus, an advance in a path of higher voltage of the gun and a low frequency is meant to increase a voltage of an accelerated beam in the described L-band EVT. Its layout does not differ from **Figure 4**. Numerical simulations have shown feasibility of reaching the energy of the accelerated beam about 2 MeV with efficiency about 60% at a gun voltage equal to 110 kV. More detailed data of the project are submitted in **Table 1**. The accelerating structure includes 12 accelerating cavities and 7 passive cavities (see **Figure 9**). The code MAGIC 2D and both models described in Section 4 have been applied for the analysis. For the results of beam dynamics of the annular model, please see **Figure 9**, too. **Figure 10** shows particle energy distribution.

7. Two-stage multi-beam linear accelerator EVT

The further increase in energy of the accelerated beam will consist in increase in a voltage of a drive beam. It is suggested to consider two-stage acceleration. Let's assume that the set of the drive beam-lets consists of two groups. At the first stage, the first group of drive beam-lets that form drive beam 1 transfer their energy to the second group of beam-lets that form drive beam 2. At the second stage, the second group of drive beam-lets transfers their energy to the accelerated beam. Such configuration is allowed to increase a transformation ratio and energy of the accelerated beam considerably. Preliminary results of numerical simulations of two-stage EVT operating in S-band with a 60 kV gun and generating a 1 A, 4.3 MV beam at its output, with an efficiency of 22%, are presented. See **Figure 12** and **Table 2**.

Figure 11 (left) shows a multi-beam cathode system. 36 peripheral cathodes generate the drive 1 beam-lets, and 6 cathodes generate the drive 2 beam-lets. The central cathode generates the accelerated beam. A rather low perveance (0.41 $\mu A/V^{3/2}$) of drive beam-lets is chosen. It is allowed to simplify the magnetic system and electron optics of the gun. The drive beam is frozen in a homogeneous magnetic field of the focusing solenoid. You can see the presence of ripples having an allowable value in **Figure 11** (right).

MAGIC 2D simulations of beam dynamics were carried out using an annular model of beams and cavities. Please see the results of numerical simulations in **Figures 12–14** and in **Table 2**.

Figure 14 shows particle energy distribution at the accelerator output. For the change in beams energy vs. distance, please see **Figure 15**.

Operating frequency	2856		MHz
Voltage of common electron gun	60		kV
Output energy of accelerated bunch	4.3		MeV
Average energy of accelerated beam	3.6		MeV
Gun current of accelerated beam	1		A
Total gun current of drive beams	252		A
Solenoid magnetic field	1000		G
Length of RF part of accelerator	1.9		m
Transmission of accelerated beam	90		%
Total efficiency of two-stage EVT	≈22		%
Stage	1st	2nd	
Gun current of drive beam	216	36	A
Perveance of drive beam-let	0.41	0.41	$\mu A/V^{3/2}$
Transformation ratio	6	36	
Voltage gain factor of stage	≈3.5	≈21	
Number of drive beam-lets	36	6	
Transmission of drive beam	98	90	%
Efficiency of stage	≈55	≈45	%

Table 2. Simulated parameters of the two-stage EVT.

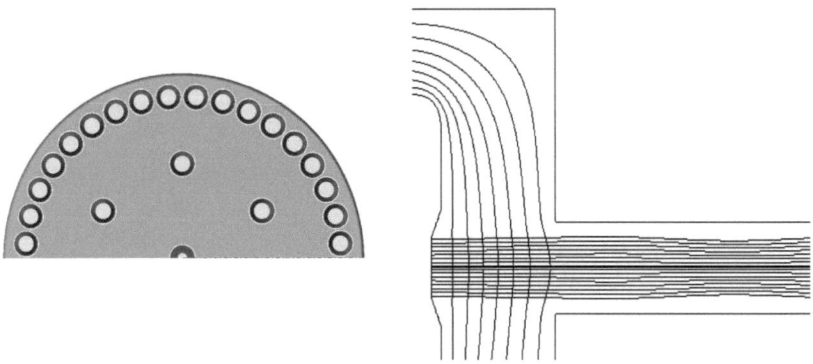

Figure 11. Multi-beam cathode system of the two-stage EVT (left) and an example of 2D simulation of the drive beam-let gun (right).

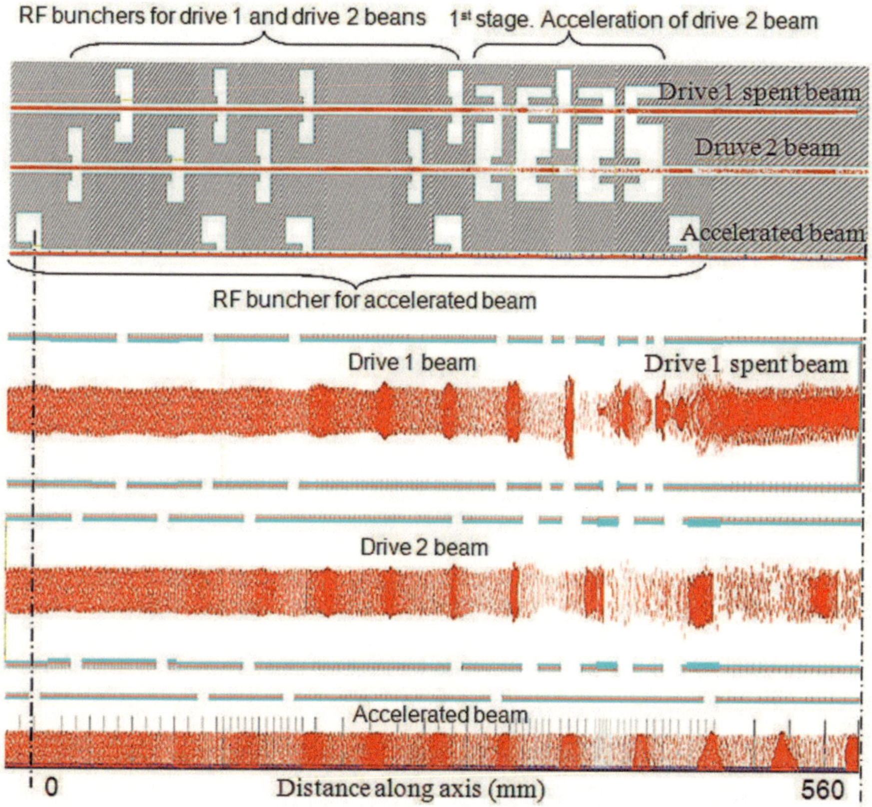

Figure 12. MAGIC 2D results of simulation of beam dynamics in the bunching part and at the 1st stage.

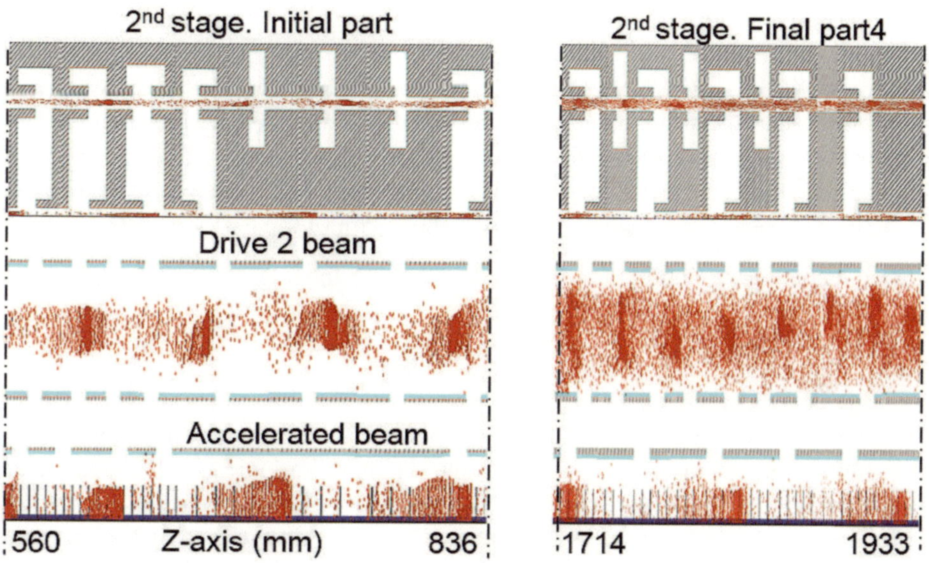

Figure 13. Accelerating structure of 2nd stage. Dynamics of drive 2 and accelerated beams.

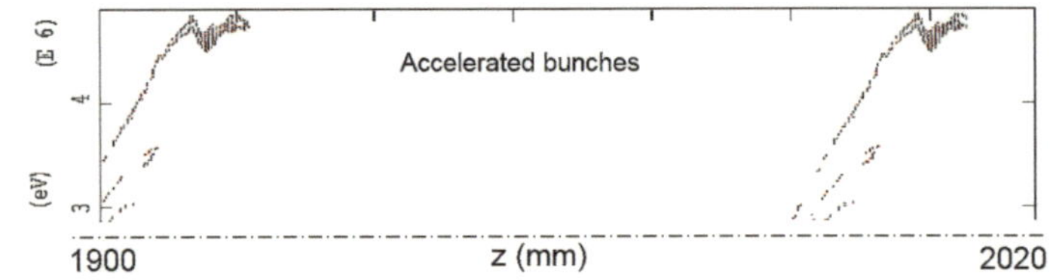

Figure 14. Energy distribution at the output of the two-stage EVT.

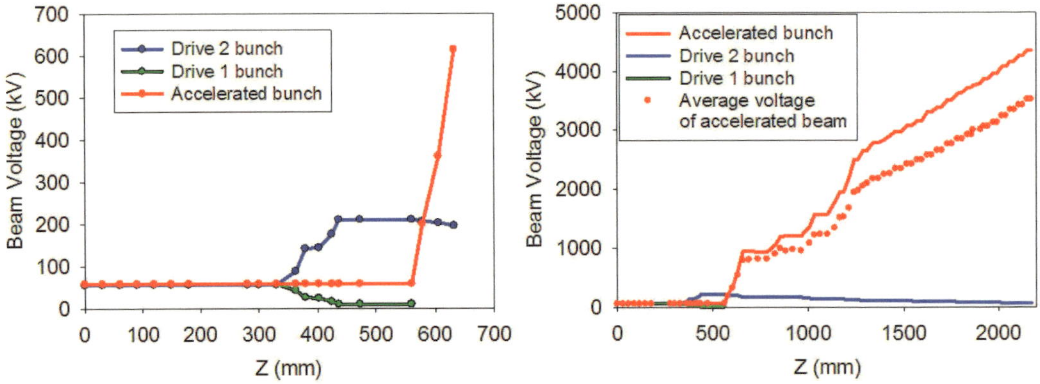

Figure 15. Change of beams voltage along the accelerator axis.

8. Depressed collector (DC) for spent drive beam

The efficiency of 66% achieved in preliminary numerical simulations of the S-band EVT is a rather high value. The further increase in efficiency of the EVT accelerators is possible at the use of a depressed collector. It is suggested considering a novel concept of a DC, which allows theoretical estimation of DC properties. The given concept is based on the thesis of existence of an extremum of the DC efficiency versus DC voltage. The concept does not struggle with the energy distribution of electrons having a place after the device output, but uses properties of energy distribution. Let us name this concept a depressed collector with a spectra separator, DC&SS. The schematic circuit of the tube with a one-stage DC is shown in **Figure 16** (left). **Table 3** shows the basic relationships describing the circuit of the EVT with a DC.

Substituting the above resulted relationships, it is derived the total tube efficiency with a DC:

$$\eta_t = \frac{\eta_e}{1 - (1 - \eta_e)\eta_c} \tag{37}$$

Eq. (36) is well known (see, for example [12]), but we shall regard it now jointly with the presence of an energy spectrum of particles leaving a tube. Variable $I_c = I_c(V_c)$ is remarkable since it is a function of a collector voltage as a consequence of the presence of energy distribution. Our subsequent problem is to study the function $\eta_c(V_c)$ in order to find an extremum at some assumptions concerning the energy distribution $j(V)$, which is normalized on the beam current I_B:

$$\int_0^\infty j(V)dV = I_B \quad (38)$$

In the gap of the separator, the beam is divided into low-voltage and high-voltage fractions. The current

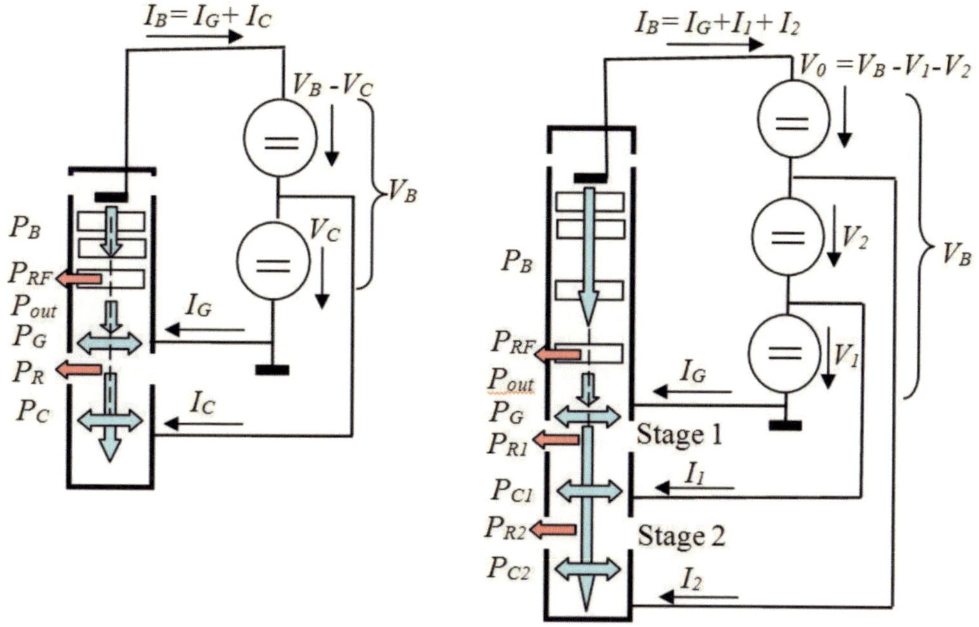

Figure 16. Schematic circuit of a tube with depressed collector. One-stage DC (left); two-stage DC (right).

Beam voltage	V_B
Voltage of stage	V_C, V_1, V_2
Current of potential part of stage	I_C, I_1, I_2
Beam power	$P_B = I_B V_B$
RF power, transfer to accelerated beam	P_{RF}
Loss power in ground part of collector	P_G
Loss power in potential part	$P_C, P_{C1}, P_{C2},$
EVT electron efficiency	$\eta_e = P_{RF}/P_B$
Beam power at output of tube	$P_{Out} = (1- \eta_e)P_B$
Recovery power	$P_R = V_C I_C$
Total power generating by high-voltage source	$P_t = V_B I_B - V_C I_C$
Efficiency of depressed collector	$\eta_c = P_R/P_{Out} = V_C I_C/P_{Out}$
Total tube efficiency with DC	$\eta_t = P_{RF}/P_t = P_{RF}/(P_B - V_C I_C)$

Table 3. Notations and relationships describing a DC.

$$I_c(V_c) = \int_{V_c}^{\infty} j(V)dV \tag{39}$$

that has reached the potential part of the collector is the fraction of the current at which $V > V_C$. In the same way, it is written down the equation for the beam power leaving the accelerating structure:

$$P_{out} = \int_{0}^{\infty} V \cdot j(V)dV \tag{40}$$

Thus, it is evaluated the fundamental formula Eq. (41) for efficiency η_c of a depressed collector included in the DC&SS concept:

$$\eta_c(V_c) = \frac{V_c \int_{V_c}^{\infty} j(V)dV}{\int_0^{\infty} V \cdot j(V)dV} \tag{41}$$

Figure 6 (right) above shows a numerical example of the energy distribution for the spent drive beam of the S-band EVT versus distance along beam axes. **Figure 17** shows the energy distribution of the same beam. The gamma distribution $x \cdot e^{-x}$ (curved line) can serve as a close approximation for numerical distribution.

Thus, let us approximate $j(V)$ by the gamma distribution:

$$j(V) = \frac{I_B}{V_0} x \cdot e^{-x} \tag{42}$$

where $x = V/V_0$, $V_0 = V_B(1 - \eta_e)/2$. Further, it is defined I_c according to Eq. (39):

$I_c(x_c) = I_B \cdot e^{-x_C}(x_C + 1)$, where $x_c = V_c/V_0$, and it is defined P_{out} according to Eq. (40):

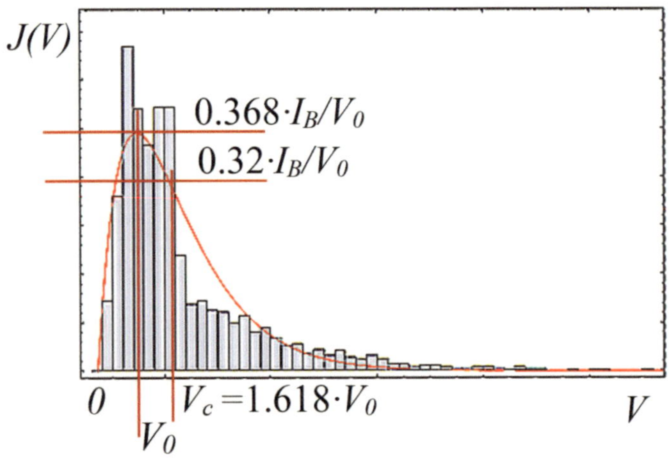

Figure 17. A numerical simulated histogram of particle distribution vs. voltage and an approximating gamma distribution are shown for the S-band EVT.

$$P_{out} = 2I_B V_0 \tag{43}$$

Substituting the found expressions in Eq. (41), it is obtained:

$$\eta_c(x_c) = \frac{1}{2} x_c e^{-x_c}(x_c + 1) \tag{44}$$

The function $\eta_c(x_c)$ has the maximum equal to **0.42** at $x_c = \frac{1+\sqrt{5}}{2} = 1.618$.

Thus, it is derived the collector efficiency and other parameters at the optimum point:

$$\eta_c = 0.42; I_c = 0,519 \cdot I_B; I_G = 0,481 \cdot I_B; V_c = 1.618 \cdot V_0. \tag{45}$$

As an example of the S-band EVT (**Table 1**), one have:

$$V_B = 6 \cdot 10^4; \eta_e = 0.66; V_0 = V_B(1 - \eta_e)/2 = 1.02 \cdot 10^4; V_c = 1.65 \cdot 10^4. \tag{46}$$

The total efficiency (according to Eq. 37) at the optimum point is $\eta_t = 0.77$, which is the upper theoretical limit!

For a DC having m stages i.e. m gaps and consequently m gap voltages (see **Figure 16**), efficiency is as follows:

$$\begin{cases} \eta_t = \dfrac{\eta_e}{1 - (1 - \eta_e)\eta_c} \\ \eta_c = \sum_{n=1}^{m} \eta_n \end{cases} \tag{47}$$

where η_c is DC efficiency in multi-stage case. The collector efficiency η_n of the nth stage is calculated according to the Eq. (41), where the lower limit of integration is equal to potential $U_n = \sum_{i=n}^{m} V_i$ of the given stage concerning the ground. This Eq. (48) defines the concept of a DC&SS in multi-stage case:

$$\eta_n(V_1, V_2, \dots V_n) = \frac{V_n \int_{U_n}^{\infty} j(V) dV}{\int_0^{\infty} V \cdot j(V) dV} \tag{48}$$

Using the approximation (see Eq. 42) for beam distribution $j(V)$, leads to:

$$\begin{cases} \eta_n = \dfrac{1}{2} \left[x_n \left(1 + \sum_{i=n}^{m} x_i \right) e^{-\sum_{i=n}^{m} x_i} \right] \\ \eta_c = \sum_{n=1}^{m} \eta_n \end{cases} \tag{49}$$

where: $x_i = V_i/V_0$, $x_n = V_n/V_0$, $V_0 = V_B(1 - \eta_e)/2$, V_i is gap voltage of the ith stage, V_B is beam voltage. For a schematic circuit of a tube with a two-stage DC, **Figure 17** (right), the total collector efficiency will be as follows:

$$\eta_c(x_1, x_2) = \eta_1 + \eta_2 = \frac{1}{2}\left[x_1(1+x_1)e^{-x_1} + x_2(1+x_1+x_2)e^{-(x_1+x_2)}\right] \quad (50)$$

The extremum of this function is equal to $\eta_c = 0.586$ at: $x_1 = 1.118$; $x_2 = 1.398$.

At $\eta_e = 0.66$, according to Eq. (37), it will correspond to the total tube efficiency $\eta_t = 0.824$.

Table 4 shows the results of efficiency η_c optimization for a collector that contains one to six stages.

m	x_1	x_2	x_3	x_4	x_5	x_6	η_c
1	1.618						0.420
2	1.118	1.398					0.586
3	0.884	0.941	1.319				0.676
4	0.743	0.732	0.876	1.276			0.734
5	0.648	0.608	0.675	0.841	1.249		0.773
6	0.578	0.526	0.558	0.645	0.819	1.229	0.803

Table 4. The optimization result for the function $\eta_c(x_1, x_2, \ldots)$ via its parameters $x_1, x_2, \ldots$.

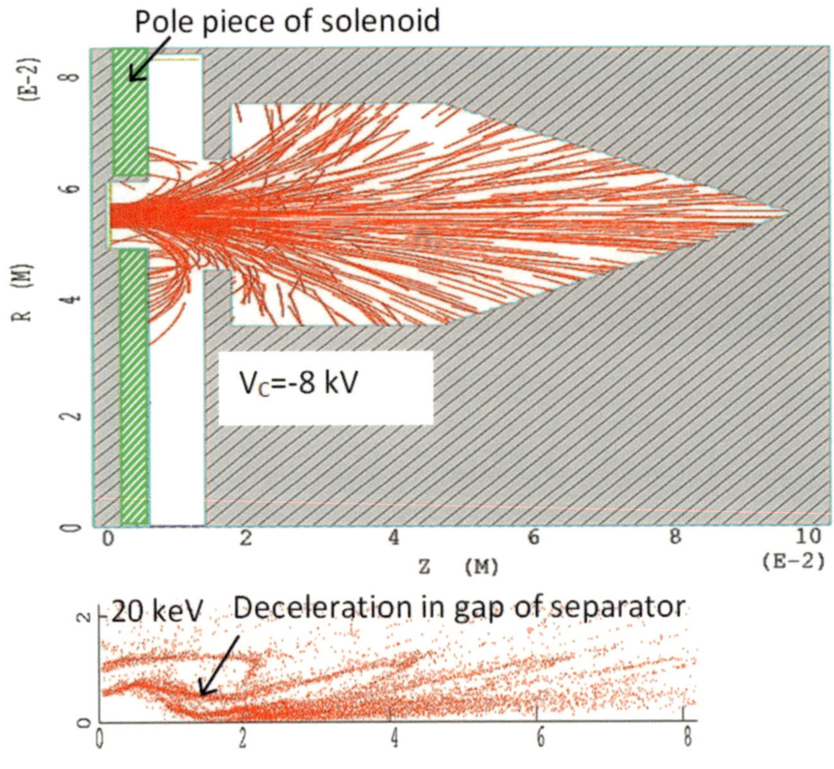

Figure 18. Geometry of a one-stage DC for the S-band EVT. A picture of particles energy is added at the bottom of the figure.

The theoretical premise of an increase in efficiency of a device with a multi-stage DC are rather optimistic. However, simulations of beam dynamics in conditions close to real do not show any noticeable increase in the device's efficiency while engineering complexity of such multi-stage DC tangibly increases. Thus, use of a one-stage DC is, perhaps, more preferable in practice.

The analytical treatment described above is rather idealized as it does not include the beam particle movement dynamics; however, it shows parameters to which there can try to approach. The result of initial simulation of a one-stage DC for a spent drive beam of the S-band EVT is shown in **Figure 18**. No special optimization of the DC geometry and the magnetic field shape was carried out. Secondary emission was not taken into consideration either. The annular 2D model of the drive beam has been used.

The results of initial simulations have shown a DC efficiency about 25%, which increases the total efficiency of the S-band EVT up to 72%. It certainly is substantially lower than the idealized theoretical results of the previous section and allows further optimization.

9. Conclusion

Theoretical and numerical research of properties of the linear accelerator EVT has shown that there is sufficient comprehension about it properties. There is no visible vagueness in technical problems. Certainly, the EVT accelerator does not show a record rate of acceleration; however, it shows that to reach a high efficiency of 70% or more is possible. It is much higher than the total efficiency of a klystron loaded with an ordinary linear accelerator. The two-stage configuration of the accelerator considerably increases the transformation ratio, but the accelerator's efficiency decreases; therefore, further optimization of this configuration is preferable. Thus, there are all preconditions in place to proceed with the engineering design.

Author details

Vladimir E. Teryaev

Address all correspondence to: vladimir_teryaev@mail.ru

Budker Institute of Nuclear Physics, Novosibirsk, Russia

References

[1] Corsini R. Two-beam linear colliders – Special issues. In: Proceedings of PAC09; 2009; Vancouver, BC, Canada. IEEE; 2009. pp. 3100-3104

[2] Derbenev Ya S, Lau YY, Gilgenbach RM. Proposal for a novel two-beam accelerators. Physical Review Letters. 1994;**72**(19):3025-3028. http://patents.justia.com/patent/5483122

[3] Teryaev VE, Hirshfield JL, Kazakov S, Yakovlev VP. Low Beam Voltage, 10 MW, L-Band Cluster Klystron. In: Proceedings of PAC09; 2009; Vancouver, Canada. http://accelconf.web.cern.ch/AccelConf/PAC2009/papers/tu5pfp093.pdf

[4] Teryaev VE, Shchelkunov SV, Kazakov SY, Hirshfield JL, Ives RL, Marsden D, Collins G, Karimov R, Jensen R. Compact low-voltage, high-power, multi-beam Klystron for ILC: Initial test results. In: Proceedings of the Division of Particles and Fields; Aug 4–8, 2015; Ann Arbor, MI, USA. http://arxiv.org/abs/1510.06065

[5] Kazakov SY, Kuzikov SV, Jiang Y, Hirshfield JL. High-gradient two-beam accelerator structure. Physical Review Special Topics - Accelerators and Beams. 2010;**13**(7):071303-071301. https://journals.aps.org/prab/abstract/10.1103/PhysRevSTAB.13.071303

[6] Dolbilov GV. Two-beam induction linear collider. In: Proceedings of EPAC 2000; 2000; Vienna, Austria. IEEE; 2000. p. 904

[7] Dolbilov GV. Two beam proton accelerator for neutron generators and electronuclear industry. In: Proceedings of the 2001 Particle Accelerator Conference; 2001; Chicago, USA. IEEE; 2001. pp. 651-653

[8] Goplen B, Ludeking L, Smithe D, Warren G. MAGIC User' Manual. MRC/WDC-R-409; 1997. DOI: http://www.orbitalatk.com/magic/default.aspx

[9] Teryaev VE, Kazakov SY, Hirshfield JL. Multi-beam linear accelerator EVT. In: Proceedings of EAAC 2015, Nucl. Instr. and Meth. in Phys. Res. A 829; Sept. 13–19, 2015; Isola Elba, Italy; 2016. pp. 221-223. https://www.researchgate.net/publication/301225661_Multi-beam_linear_accelerator_EVT

[10] Teryaev V. Influence of Shape of a Charged Bunch on it's Interaction with the Cavity. KEK Report 2008–9; 2009. Available from: https://www.researchgate.net/publication/265849229_INFLUENCE_OF_SHAPE_OF_A_CHARGED_BUNCH_ON_IT%27S_INTERACTION_WITH_THE_CAVITY_The_engineering_approach

[11] Deruyter H, Mishin A, Roumbanis T, Schonberg R, Farrell S, Smith R, Miller R. High power 2 MeV linear accelerator design characteristics. In: Proceedings of EPAC96; 1996; IEEE; 1996. Available from: http://accelconf.web.cern.ch/accelconf/e96/PAPERS/THPG/THP119G.PDF

[12] Wenlong H, Donaldson CR, Zhang L, Ronald K, Phelps ADR, Cross AW. Numerical simulation of a Gyro-BWO with a helically corrugated interaction region, cusp electron gun and depressed collector. In: Prof. Jan Awrejcewicz, editor. Numerical Simulations of Physical and Engineering Processes. ISBN: 978–953–307-620-1 ed. InTech; 2011. pp. 101-131. https://www.intechopen.com/books/numerical-simulations-of-physical-and-engineering-processes/numerical-simulation-of-a-gyro-bwo-with-a-helically-corrugated-interaction-region-cusp-electron-gun

Chapter 4

Optically Controlled Laser-Plasma Electron Acceleration for Compact γ-Ray Sources

Serge Y. Kalmykov, Xavier Davoine,
Isaac Ghebregziabher and Bradley A. Shadwick

Additional information is available at the end of the chapter

http://dx.doi.org/10.5772/intechopen.71679

Abstract

Thomson scattering (TS) from electron beams produced in laser-plasma accelerators may generate femtosecond pulses of quasi-monochromatic, multi-MeV photons. Scaling laws suggest that reaching the necessary GeV electron energy, with a percent-scale energy spread and five-dimensional brightness over 10^{16} A/m^2, requires acceleration in centimeter-length, tenuous plasmas ($n_0 \sim 10^{17}$ cm^{-3}), with petawatt-class lasers. Ultrahigh per-pulse power mandates single-shot operation, frustrating applications dependent on dosage. To generate high-quality near-GeV beams at a manageable average power (thus affording kHz repetition rate), we propose acceleration in a cavity of electron density, driven with an incoherent stack of sub-Joule laser pulses through a millimeter-length, dense plasma ($n_0 \sim 10^{19}$ cm^{-3}). Blue-shifting one stack component by a considerable fraction of the carrier frequency compensates for the frequency red shift imparted by the wake. This avoids catastrophic self-compression of the optical driver and suppresses expansion of the accelerating cavity, avoiding accumulation of a massive low-energy background. In addition, the energy gain doubles compared to the predictions of scaling laws. Head-on collision of the resulting ultrabright beams with another optical pulse produces, via TS, gigawatt γ-ray pulses having a sub-20% bandwidth, over 10^6 photons in a microsteradian observation cone, and the observation cone, and the mean energy tunable up to 16 MeV.

Keywords: laser wakefield acceleration, optical control of injection, optical shock, negative chirp, pulse stacking, Thomson scattering, particle-in-cell simulations

1. Introduction

Particle accelerators are among the largest and most expensive scientific instruments. Their large footprint is dictated by the modest acceleration gradient (in tens of MeV per meter), limited by the breakdown of metallic accelerating cavities. Accelerating electrons in the fully or partially ionized medium (i.e. a plasma) lifts this limitation, making accelerators thousands of times smaller, literally "table-top." Since plasmas are free of the damage limits of conventional accelerators, they may build up TV/m fields within structures propagating at a near-luminal speed. First ideas of harnessing collective plasma fields to actively control the phase space of a high-energy electron beam (e-beam), were brought into the world over 60 years ago [1–3]. Yet, it was not until the first decade of this century that the accelerator community started witnessing systematic progress in plasma acceleration of electron and positron beams [4–8].

Competition with conventional linear accelerators in generation of quasi-monoenergetic (QME) e-beams requires independently driven near-luminal, high-field plasma structures, such as Langmuir plasma waves [9–11] or cavities ("bubbles") of electron density [12–15]. The accelerating buckets must retain their shape in the course of propagation or change the shape and potentials in a controllable fashion to avoid degradation of the externally injected e-beam. It is equally important for the injection mechanism to ensure subsequent acceleration of the beam without picking up additional unwanted charge (the "dark current"). To this end, control over driver evolution and the plasma density profile is of paramount importance.

Driving the accelerating plasma structures with a radiation pressure of a femtosecond, multi-terawatt (TW) laser pulse (hence the term "laser wakefield") provides abundant opportunities for all-optical control of both injection and acceleration processes [16, 17]. Early demonstrations of QME laser-plasma acceleration [4–6] were a perfect example of this control. It was not until the optical driver closely matched the plasma parameters, to ensure its propagation as a whole, without breaking up longitudinally or transversely, that the long coveted QME electron bunches were realized. The matching [18] made it possible for the laser to produce a "bubble" almost completely devoid of electrons in its immediate wake [19, 20]. The bubble acts at the same time as a nonlinear waveguide for the laser pulse and an accelerating bucket for the electrons. The ponderomotive force of the pulse maintains the bubble shape. It expels all electrons facing the pulse (hence the term "blowout regime"), while the bulk electrons are attracted to the propagation axis. The difference between attractive force due to the charge separation and the repulsive radial ponderomotive force controls the trajectories of electrons making up the bubble shell. The resulting soft channel, approximately replicating the three-dimensional (3-D) shape of the pulse [19], evolves in a lock-step with the optical driver [15, 16, 20–25]. In consequence, it traps initially quiescent background electrons, eliminating the need for an external photocathode [16, 21, 22]. Notably, in the regimes featuring production of low-emittance e-beams, only a tiny minority of electrons making up the bubble shell are trapped and subsequently accelerated. Their collective fields, i.e., beam loading [26], contribute very little to the bubble evolution and are unable to change the kinetics of self-injection [21–23].

Two fundamental relativistic optical phenomena underpin the matching conditions [18]. The first one is relativistic self-focusing. As electrons oscillate in the field of focused laser beam, the relativistic increase in their mass and, hence, the nonlinear refractive index reach maximum near axis, where the laser intensity is the highest. The plasma thus acts as a focusing fiber, compensating for diffraction. If the pulse power P exceeds the critical value $P_{cr} = 16.2(n_c/n_0)$ GW [27] even by a few percent, the self-focusing will saturate [28] only at the point of full electron blowout [13]. Here, n_0 is the background electron density, $n_c = m_e\omega_0^2/(4\pi e^2)$ is the critical density for radiation with a frequency ω_0, and m_e and $-|e|$ are the electron rest mass and charge. Matching the spot size of the incident pulse to the value

$$r_\mathrm{m} = 2^{3/2}k_p^{-1}(P/P_\mathrm{cr})^{1/6} \tag{1}$$

balances the force due to the charge separation and the ponderomotive force acting upon the electron at the boundary of the bubble. Here, $k_\mathrm{p} = \omega_\mathrm{pe}/c = 1.88\sqrt{n_{20}}\ \mu\mathrm{m}^{-1}$ is the plasma wave number, $\omega_\mathrm{pe} = \sqrt{4\pi e^2 n_0/m_e} \ll \omega_0$ is the Langmuir plasma frequency, c is the speed of light in vacuum, and n_{20} is the background density in units $10^{20}\ \mathrm{cm}^{-3}$. The matched pulse propagates in a single filament confined to the bubble. Conversely, strong mismatching results in a transverse breakup of the pulse, massive energy loss to the plasma, and disruption of self-guiding [29, 30]. The other key physical phenomenon, which limits electron energy gain, is self-phase-modulation, viz. accumulation of frequency red shift, imparted by the wake, at the self-phase-modulation. As the pulse propagates, this shift reaches a large fraction of ω_0, while the negative group velocity dispersion (GVD) in the plasma delays these low-frequency components, etching away the pulse leading edge. In the frame of reference comoving with the bubble, these components start to accumulate around the point where electron density drops to zero[1], building up an optical shock with a subcycle rising edge [16, 22–25]. If the bubble were nonevolving, the etching velocity would be its phase velocity, which defines the electron dephasing length, $L_d = (2/3)(n_c/n_0)r_\mathrm{m}$ [18]. In addition, etching velocity defines the pulse energy loss, which also limits electron energy gain. The pulse loses most of its energy and is unable to drive the bubble after a distance $L_\mathrm{depl} = (n_c/n_0)c\tau_L$ (the depletion length), where τ_L is the duration of the incident pulse. Matching the dephasing and depletion lengths, so that $\tau_L = 2r_\mathrm{m}/(3c)$, promises to maximize the acceleration efficiency and, possibly, reduce electron energy spread via phase space rotation at the end of acceleration cycle. Under the matching condition, the maximal energy gain scales as [18]

$$\Delta E[\mathrm{GeV}] \sim 0.125(P[\mathrm{PW}])^{1/3}\left(n_{20}\lambda_{0,\,\mu\mathrm{m}}^2\right)^{-2/3} \tag{2}$$

Here, $\lambda_{0,\,\mu\mathrm{m}}$ is the laser pulse wavelength, $\lambda_0 = 2\pi c/\omega_0$, in microns. To ensure robust self-guiding and preserve self-injection, the power ratio must be at least $P/P_\mathrm{cr} \equiv \kappa > 10$, or $n_{20}\lambda_{0,\,\mu\mathrm{m}}^2 > 1.8 \times 10^{-4}\kappa(P[\mathrm{PW}])^{-1} > 1.8 \times 10^{-3}(P[\mathrm{PW}])^{-1}$. This, in combination with (2), yields a rather discouraging scaling,

$$\Delta E[\mathrm{GeV}] < 40\kappa^{-2/3}P[\mathrm{PW}] < 8.6P[\mathrm{PW}] \tag{3}$$

According to (3), GeV energy gain in the matched regime requires at least 117 TW laser power, or 3.75 J per 32 fs matched pulse. The matched plasma density is, in this case, $n_0 = 2.4 \times 10^{18}\ \mathrm{cm}^{-3}$,

[1]If electron evacuation is incomplete, and the frequency red shift is significant, the mid-IR radiation may further slide into the bubble, building inside it a single-cycle mid-IR pulse [31, 32] or even another optical shock [33, 34].

and the dephasing/depletion length is $L_d = 6.9$ mm. Laboratory experiments with cm-length gas jets from slit nozzles had approached this regime very closely, demonstrating background-free e-beams with the energy up to 900 MeV, yet at the repetition rate below 10 Hz [35]. Conversely, generating these near-GeV e-beams at a kHz repetition rate, for the applications dependent on dosage, would call for a 4 kW average-power laser amplifier, a technology of the distant future [36, 37]. Evidently, existing sub–50 TW systems are limited to the modest sub-450 MeV yields.

Apart from frustrating the production of GeV beams at a high repetition rate, this matching strategy only partly solves the problem of e-beam quality. While aiming to stabilize transverse dynamics and avoid filamentation of the drive pulse, the physical arguments leading to the scaling (2) assume that *the pulse self-compression remains unaltered.* Yet this process, apart from limiting the energy gain, destroys e-beam most assuredly if acceleration extends through the pulse depletion. (A plethora of evidence exists to this effect, both in laboratory experiments and numerical simulations [16, 19, 22–25, 38–42].) To enable a new generation of compact particle and radiation sources [43, 44], one has to bypass the limitations this of scaling by designing an optical driver resilient to self-phase-modulation and self-compression. Photon engineering of this kind, aiming to produce e-beams capable to emit quasi-monochromatic, high-flux γ-ray pulses via Thomson (or inverse Compton) scattering [44], is the focus of this chapter.

Inverse Compton scattering is an emerging radiation generation technique [25, 44–52], which has already shown its potential for obtaining quasi-monochromatic, strongly collimated γ-ray pulses through the collision of a short QME e-beam and a mid-IR to UV interaction laser pulse (ILP) [53–73]. During the interaction, relativistic electrons, propagating at an angle to the ILP, experience its Lorentz-compressed wave front, the maximum compression occurring along the e-beam direction. As they oscillate in the ILP electromagnetic field, electrons emit radiation, scattering the compressed wave front. An observer in the far field thus detects an angular distribution of high-energy photons, with the energy being the highest for a detector placed in the e-beam direction. For the head-on collision, the ILP photon energy is Doppler up-shifted by a factor $4\gamma_e^2$, where γ_e is the electron Lorentz factor. A beam of 900 MeV electrons thus converts 1.5 eV ILP photons into 19 MeV γ-photons. As the energy of emitted photons is much lower than the electron energy, the recoil is negligible. This low-energy semi-classical limit of the general quantum-mechanical inverse Compton scattering, known as Thomson scattering (TS), is the subject of this chapter. E-beams from conventional accelerators [53–62], produce multipicosecond TS γ-ray pulses. These have a high degree of polarization and are thus attractive as e-beam diagnostics [53, 54]. Their other applications are generation of polarized positrons from dense targets [55] and nuclear resonance fluorescence studies [56–61]. However, the large footprint of conventional accelerators makes such radiation sources scarce and busy user facilities. In addition, the large (cm-scale) size of the radio-frequency–powered acceleration cavities makes it difficult to generate and synchronize e-beams (and, hence, TS γ-ray pulses) on a subpicosecond time scale relevant to high-energy density physics [74]. Luckily, a miniature LPA offers an alternative technical solution that permits production of even shorter (viz. femtosecond), yet high-current (viz. kA) e-beams [75]. To drive narrowband TS γ-ray sources, these beams have to meet some minimal requirements, such as a combination of a near-GeV energy with a percent-scale energy spread, a five-dimensional (5-D) brightness above 10^{16} A/m^2 [76], and preferably absent low-energy background. These requirements, in combination with the kHz-scale repetition rate dictated by the applications, are clearly conflicting even for the most ambitious laser technology [77, 78]. LPA experiments, guided by the theoretical scaling (2), are presently struggling to reach this level of performance. Typically, acceleration through pulse depletion, carried out in pursuit of ever higher energy, consistently builds up massive energy tails in the e-beams [39–42]. These beams produce a large-bandwidth γ-ray TS signal [63–73], which is incompatible with applications in nuclear photonics and radiography [48, 49, 60]. The current trend is to use the existing low-quality beams and try extending the high-energy tail of the photon distribution beyond 10 MeV, by using higher harmonics of the ILP [72], or by using few-GeV electrons from single-shot petawatt LPA facilities [73], or by employing an ILP of relativistic intensity [68].

Seeking the remedy to this situation, we take advantage of the fact that the LPA e-beams readily lend themselves to all-optical manipulation. Modifying the drive pulse dynamics, through a judicious choice of its phase and shape, alters kinetics of electron self-injection. This, in turn, introduces modulations to the e-beam current and/or imparts a chirp to its longitudinal momentum. As the e-beam phase space imprints itself onto the spectrum of emitted photons, all-optical control of electron source enables tailoring the TS γ-ray signal [25, 52].

As explained earlier, the plasma response compresses the optical driver of a conventional LPA (i.e. a transform-limited multi-TW pulse) into a subcycle relativistic optical shock; this happens long before electron dephasing. The shock snowplows the ambient plasma electrons, causing electron density pileup inside the shock and a multifold increase in the field of charge separation behind it [16, 23]. The resulting uncontrolled elongation of the bubble causes massive continuous injection of electrons from its shell. Because of this dark current, caused by the uncompensated adverse optical process, maximization of the energy gain conflicts with the preservation of e-beam quality. We propose to resolve this conflict by incoherently mixing the pulse at the fundamental frequency with a frequency-upshifted pulse of the same, or lower, energy (on a sub-Joule scale) [24, 52]. As the photon diffusion rate due to GVD drops as the frequency grows, the blue-shifted stack component is resilient to self-compression. Because of the strong frequency dependence of the diffusion rate ($\sim \omega^{-3}$) [79], even a modest 25% frequency up-shift appears to be sufficient [24]. Simulations show that even the stack of fully overlapping components, in the fashion of Ref. [80], remains resilient to self-compression (at least on the time scale of electron dephasing). The presence of the almost undeformable blue component does not permit formation of the intensity gradients at the subcycle scale. In the absence of the optical shock, the bubble expansion and, hence the dark current, is insignificant. The particle flux and charge in the energy tail drop multifold in comparison to the reference case of a transform-limited optical driver (the latter complying fully with the scaling (2)). Advancing the blue-shifted component by $T \sim \tau_L$ improves the situation even further. Emulating, in this way, a piecewise, large-bandwidth negative chirp, we essentially place a protective screen ahead of the vulnerable unshifted tail. As this "hard hat" plows through the plasma, expelling background electrons, the soft tail maintains the bubble shape, thus defining kinetics of self-injection. In the regime of our simulations, the dephasing length, defined by the etching of the head, extends by almost 80%, while the electron energy doubles against the reference case, reaching almost 900 MeV over 2.5 mm acceleration distance. Regardless of the time delay, very quiet injection keeps the e-beam brightness above 4×10^{16} A/m^2, favoring the use of these beams in Thomson sources [76]. In the case of a time-delayed stack, extracting the e-beam before dephasing (using, for instance, a gas cell target of variable length [81]), thus changing the e-beam energy in the interval 400–900 MeV, preserves its 10^{17} A/m^2 brightness. This permits the tuning of mean energy of the TS γ-ray signal between 4 and 16 MeV, preserving 1.5×10^6 photons in a microsteradian observation solid angle. Notably, the low energy in the stacked driver (1.4 J) and the ILP (25 mJ) permits maintaining a half-kHz repetition rate while staying below kW average power, a hard yet practical task [77]. A longer ILP would help increase the photon yield by another order of magnitude, without jeopardizing the repetition rate. Overall, this brings an expectation of greater than 10^9 ph/s yield, which is not as high as 10^{13} ph/s permitted by large linacs [57], yet sufficient to identify considerable masses of enriched uranium within minutes [61]. From the viewpoint of laboratory practice, computerized manipulations of the phase and shape of the sub-Joule stack components, using adaptive optics and genetic algorithms [82, 83], should aid greatly in practical realization of the system. This optimization approach is especially effective at a kHz-scale repetition rate and low pulse energy.

The structure of the chapter is as follows: Section 2 describes the computational approach and defines parameters of the case studies. These parameters are representative of LPA experiments carried out in numerous laboratories worldwide. The reported case studies may thus serve as a reference for practical realization of the scheme in an existing experimental setting. In Section 3, we demonstrate the efficiency of using the stacks for generation of low-background electron beams, doubling their energy in comparison with the predictions of accepted scaling laws. Section 4 explores all-optical control over parameters of QME e-bunches, through variation of the time delay between the stack components. A 60% increase in electron energy and a factor 3.5 increase in brightness are demonstrated using the same target and the same energy in the stack. This permits the tuning of the TS γ-ray pulse parameters (mean energy, photon yield, and power) in the broad range of interest for nuclear photonics applications [49]. Section 5 summarizes the results and points out directions for future work.

2. Interaction regimes and simulation methods

Reduced and full 3-D particle-in-cell (PIC) simulations shed light on the physical processes essential for e-beam shaping. Quasistatic simulations using the relativistic, cylindrically symmetric, optical cycle-averaged code WAKE [13, 28] associate the massive dark current with the transformation of the optical driver into an optical shock and help develop a strategy for dark current reduction. WAKE computes the complex envelope of the laser vector potential using an extended paraxial solver. The solver preserves GVD in the presence of large frequency shifts and accurately calculates radiation absorption due to wake excitation [32, 84]. The grid $\Delta\xi \approx \Delta r/3 \approx \lambda_0/13 \approx 63$ nm and time step $\Delta t \approx 1.325/\omega_0$, with 30 macroparticles per radial cell, are sufficient to capture all physics relevant to pulse propagation and evolution of the bubble. Here, $\xi = z - ct$ and $r^2 = x^2 + y^2$. WAKE includes 3-D test particle tracking in the full (unaveraged) electromagnetic fields. Test-particle simulations allow the study of the physical process of self-injection (bubble and driver evolution) in the absence of effects due to beam loading [21–23]. Simulations using the relativistic, fully explicit, quasicylindrical code CALDER-Circ [85] explore manipulations of e-beam phase space, leading to production of clean and tunable beams. The code uses a numerical Cherenkov-free electromagnetic solver [86] and third-order splines for the macroparticles. These features, in combination with a fine grid $\Delta z = 0.125c/\omega_0 \approx 16$ nm $\approx \Delta r/16$, small time step $\Delta t = 0.1244/\omega_0$, and 45 macroparticles per cell, maintain low sampling noise and negligible numerical dispersion and avoid numerical emittance growth.

We demonstrate the limits of all-optical control fixing the total laser energy at 1.4 J. This energy may be concentrated in a single, transform-limited, linearly polarized Gaussian pulse with a carrier wavelength $\lambda_0 = 0.805$ μm and full width at half-maximum in intensity $\tau_L = 20$ fs (the reference case). The plasma begins at $z = 0$ with a 0.5 mm linear ramp, followed by a uniform section of density $n_0 = 6.5 \times 10^{18}$ cm^{-3}. The pulse, propagating toward positive z, is focused at the plasma border into a spot $r_0 = 13.6$ μm. The electric field in the focal plane is

$$\mathbf{E}_0(x, y, z = 0, t) = \mathbf{e}_x(m_e\omega_0 c/|e|)\mathcal{E}_0 \exp\left(-i\omega_0 t - 2\ln 2t^2/\tau_L^2 - r^2/r_0^2\right) \quad (4)$$

Here, $\mathbf{e}_x$ is the unit polarization vector, and the normalization factor $m_e\omega_0 c/|e| = 4$ TV/m. The single 70 TW pulse ($\mathcal{E}_0 = 3.27$) has $\tau_L = 2r_{\mathrm{m}}/(3c)$ and thus depletes at the point of electron dephasing, promising to maximize acceleration efficiency. Contrary to expectations, this strategy leads to copious dark current and overall low beam quality. Thus we seek to compensate for the red shift imparted by the wake. This task requires a very broadband negative frequency chirp, that is, blue shifting the leading edge by an amount comparable to the carrier frequency [23–25]. As the required broad-bandwidth, kW average-power laser amplifiers are not going to be available soon [37], we propose to synthesize a large-bandwidth, negative piecewise chirp, by optically mixing transform limited, narrow-bandwidth blocks of the same or different energy [24]. We demonstrate the emerging opportunities by splitting the 1.4 J energy evenly between two linearly (orthogonally) polarized, transform-limited 20 fs pulses, one of which, as shown in **Figure 1(b.1)** and **(c.1)**, is significantly blue-shifted and may be advanced in time. Electric field of this incoherent stack is $\mathbf{E}_{\text{stack}}(z = 0) = \mathbf{E}_0 + \mathbf{E}_{\text{head}}$, with

$$\mathbf{E}_{\text{head}} = \mathbf{e}_y(m_e\omega_0 c/|e|)\mathcal{E}_{\text{head}} \exp\left(-i\omega_{\text{head}}(t+T) - 2\ln 2(t+T)^2/\tau_L^2 - r^2/r_0^2\right) \quad (5)$$

Here, $\mathcal{E}_{\text{head}} = \mathcal{E}_0 = 2.31$, $\mathbf{e}_y$ is the unit polarization vector, the delay T is positive, and $\Omega = \omega_{\text{head}}/\omega_0 > 1$. Changing the frequency ratio and the delay permits accessing a broad range of e-beam parameters. We demonstrate the limits of this range by setting $\Omega = 1.5$. (Ref. [24] reports on other options, with equally promising results.) We consider the case of full overlap, $T = 0$ (case A), and then introduce the time delay that maximizes electron energy, $T = 3\tau_L/4 = 15$ fs (case B). **Figure 1** shows that both stacks shrink slowly compared to the reference case, while advancing $\mathbf{E}_{\text{head}}$ in time further increases rigidity of the stack.

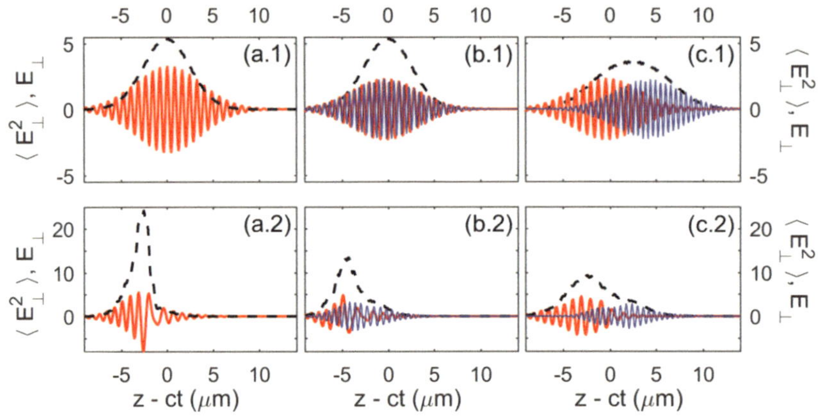

Figure 1. Snapshots of electric field on axis (in units of $m_e\omega_0 c/|e| = 4$ TV/m) show that the frequency shift between stack components makes the stack resilient to self-compression, while the nonzero time delay further increases this resilience. The pulses propagate to the right; $z = ct$ is the centroid of the carrier-frequency component (E_0) in vacuum. Red (thick dark gray): E_0. Blue (thin gray): E_{head} (in the simulation, $\mathbf{E}_0 \perp \mathbf{E}_{\text{head}}$). Dashed curve: $\langle E_\perp^2 \rangle = \langle E_0^2 \rangle + \langle E_{\text{head}}^2 \rangle$, where $\langle \cdots \rangle$ denotes averaging over an optical cycle. Panels (a) correspond to the reference case, (b) to case A ($T = 0, \Omega = 1.5$), and (c) to case B ($T = 15$ fs, $\Omega = 1.5$). Top row shows the fields at the plasma entrance ($z = 0$). Bottom row corresponds to (a.2) $z = 1.6$ mm (point of full compression in the reference case), (b.2) and (c.2) $z = 2.15$ mm (dephasing in case A).

To simulate the TS [51], we extract N_b macroparticles from the first and second buckets of the wake and use them to sample the six-dimensional (6-D) phase space of the e-beam. We then propagate a corresponding distribution of non-interacting electrons in free space by solving the relativistic equations of motion. In the absence of a laser field, the trajectories are ballistic. The e-beam collides head-on with the linearly polarized (in the x-direction) interaction laser pulse, which we specify analytically in the paraxial approximation. The ILP has a 0.8 μm carrier wavelength (photon energy $E_{\text{int}} = 1.55$ eV), 250 fs duration corresponding to 0.3% FWHM bandwidth in spectral intensity, and a 16.8 μm waist size (corresponding to a Rayleigh length of 1.1 mm). Timing between the e-beam and the ILP is such that the centroid of the beam and the peak of the ILP intensity arrive at the ILP focal plane simultaneously. Since in all regimes under consideration the e-beams are relativistic and low-density, $n_e \langle \gamma_e \rangle^{-3} \ll 10^{16}$ cm^{-3}, space charge forces are negligible [46, 47]. As the energy radiated by an electron passing through the ILP is small if compared to the energy of the electron, the recoil and radiation damping are also negligible. The ILP is shorter than 7% of its Rayleigh length and the e-beam spot size is in the submicron range; hence, the interaction occurs in an almost plane-wave geometry. To avoid broadening the TS spectra [44, 50, 68], a linear interaction regime is chosen, with the ILP normalized vector potential $a_{\text{int}} = 0.1$ (hence the ILP energy 25.5 mJ). Using the computed orbits of individual electrons and taking a weighted average over the ensemble yield the mean energy density radiated per unit frequency ω and solid angle Ω per electron [87]:

$$\frac{d^2 I_e}{d\omega d\Omega} = \frac{e^2\omega^2}{4\pi^2 c}\left(\sum_{i=1}^{N_b} w_i\right)^{-1} \sum_{i=1}^{N_b} w_i \left| \int_{-\infty}^{\infty} \mathbf{n} \times (\mathbf{n} \times \boldsymbol{\beta}_i) \exp\left(i\omega(t - \mathbf{n}\cdot\mathbf{r}_i(t)/c)\right) dt \right|^2 \quad (6)$$

Here, w_i is the macroparticle weight; $\mathbf{n}$ is the unit vector in the direction of observation; and $\mathbf{r}_i$ and $\boldsymbol{\beta}_i = \mathbf{v}_i/c$ are the radius vector and normalized velocity of the electron. A beam with a charge Q radiates the energy $d^2I_{\text{tot}}/d\omega d\Omega = (Q/|e|)d^2I_e/d\omega d\Omega$. In all cases, we show the TS spectra for the emission in the direction of e-beam propagation (on-axis observation).

3. Stacking suppresses dark current

Propagation dynamics of a bi-color stack is entirely different from that of the quasi-monochromatic reference pulse. To track changes in the pulse evolution brought about by the stacking, we use the laser vector potential, $\tilde{a} = a(r, z, \zeta)e^{-i\omega_0\zeta}$, where $\zeta = \xi/c$ is a retarded time, and $a(r, z, \zeta)$ is the complex envelope from WAKE simulations. **Figure 2(b)** and **2(c)** show, for the reference pulse and stack B, the radially integrated mean frequency and frequency variance [88],

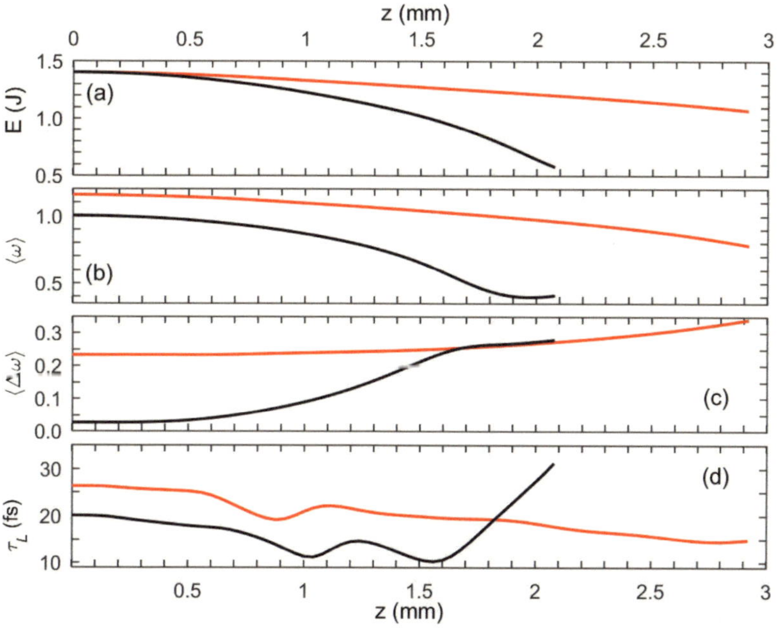

Figure 2. Negative chirp synthesized via pulse stacking mitigates frequency red shift and slows down self-compression (WAKE simulations). The pulses propagate in a uniform plasma toward positive z. Evolution of the (a) pulse energy; (b) mean frequency; (c) frequency variance; and (d) pulse length computed from the ζ-variance of the energy density on axis is shown in the reference case (black) and in case B (red/dark gray). The frequency is in units of ω_0. The curves terminate as soon as electrons reach dephasing. The negative chirp reduces depletion, red shift, and spectral broadening, thus mitigating contraction of the pulse. Using the stacked driver with a blue component advanced in time (case B) strongly delays dephasing.

$$\langle\omega(z)\rangle = A^{-1}\int_0^\infty r\mathrm{d}r\omega|\tilde{a}(r,z,\omega)|^2\mathrm{d}\omega,$$
$$\left\langle\Delta\omega(z)^2\right\rangle = A^{-1}\int_0^\infty r\mathrm{d}r\int_0^\infty(\omega-\langle\omega\rangle)^2|\tilde{a}(r,z.\omega)|^2\mathrm{d}\omega. \tag{7}$$

Here, $\tilde{a}(r,z,\omega) = \int_{-\infty}^{+\infty}\tilde{a}(r,z,\zeta)\mathrm{e}^{-\mathrm{i}\omega\zeta}\mathrm{d}\zeta$ is the Fourier transform of the laser vector potential and $A(z) = \int_0^\infty r\mathrm{d}r\int_0^\infty|\tilde{a}(r,z,\omega)|^2\mathrm{d}\omega$. **Figure 2(d)** shows the mean pulse length computed from the ζ-variance of intensity on axis,

$$\tau_L^2(z) = 8\ln 2B^{-1}\int_{-\infty}^{+\infty}(\zeta-\langle\tau\rangle)^2|a(0,z,\zeta)|^2\mathrm{d}\zeta \tag{8}$$

Here, $\langle\tau(z)\rangle = B^{-1}\int_{-\infty}^{+\infty}\zeta|a(0,z,\zeta)|^2\mathrm{d}\zeta$ is the position of the pulse centroid and $B(z) = \int_{-\infty}^{+\infty}|a(0,z,\zeta)|^2\mathrm{d}\zeta$. **Figure 3** links the local frequency shift to the longitudinal distortion of the pulse. The frequency shift is extracted from the phase of the normalized vector potential, $\tilde{a}(0,z,\zeta) = |a|\mathrm{e}^{-\mathrm{i}\omega_0\zeta+\mathrm{i}\phi}$, using two independent methods [84]. First, the Wigner transform

$$W(\xi,\omega) = \int_{-\infty}^{+\infty}\tilde{a}^*\left(\zeta+\frac{\zeta'}{2},z\right)\tilde{a}\left(\zeta-\frac{\zeta'}{2}\right)e^{i\omega\zeta'}\frac{d\zeta'}{2\pi} \tag{9}$$

yields variation of the "photon density" in time at a given point z on axis. Second, we calculate the instantaneous frequency using the rate of the envelope phase change, $\omega(\zeta) = \omega_0 - d\phi/dt = \omega_0 - \partial\phi/\partial\zeta$. Mean frequency, frequency variance, pulse duration, and photon density are experimentally measurable markers of the nonlinear optical processes. They help identify the regimes of pulse propagation and wakefield excitation [89].

Figure 3(a.1) and **3(b.1)** reveal destruction of the reference pulse at the point of electron dephasing. As the pulse plows through the plasma, it maintains the comoving negative gradient in the nonlinear index of refraction, located at its leading edge [16, 23]. A large local frequency red shift gradually accumulates along the index gradient, eventually exceeding $\omega_0/2$. **Figure 2(a)**−**(c)** show that, in the process, the energy and mean frequency of the reference pulse drop by 60%, while the mean bandwidth increases 10-fold. The negative GVD of the plasma slows down the red-shifted radiation components, etching away the pulse leading edge. One can clearly see, in the top inset in **Figure 3(b.1)**, the resulting cycle-length optical shock of relativistic intensity. What is more, the photons making up the optical shock keep sliding into the bubble, filling it with the mid-IR radiation at $\omega < \omega_0/4$. Mixing radiation of different frequencies and uncorrelated phases leads to sharp variations in the envelope phase, making the local frequency poorly defined, causing oscillations of the envelope in the tail area. More importantly, the reference pulse fully contracts long before electron dephasing (from 20

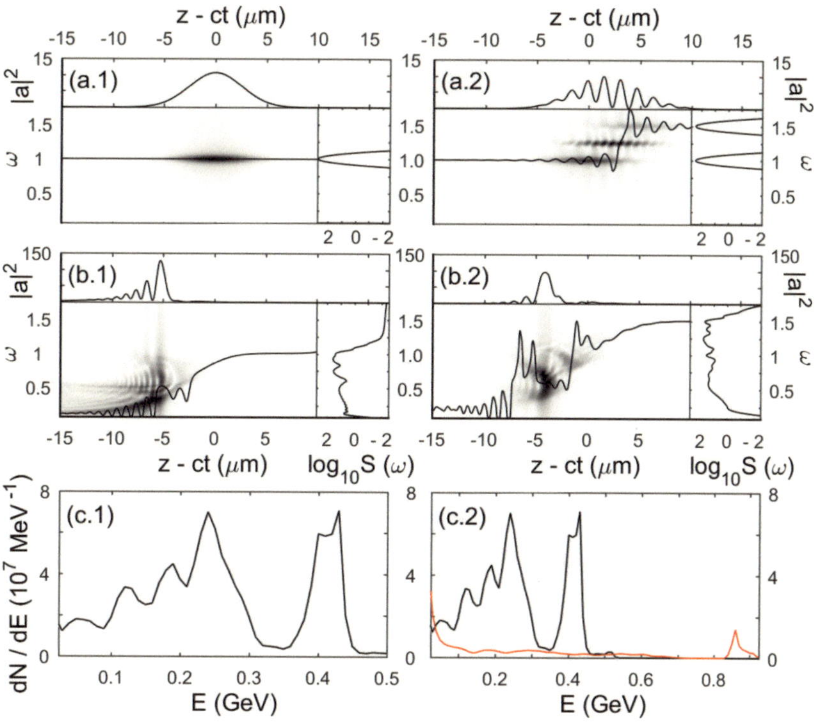

Figure 3. Deformations of the optical driver dictate e-beam quality. Increasing resilience of the driver to self-compression, via negative piecewise chirp imparted by stacking, suppresses the low-energy tail. Left column: Reference case. Right column: Case B ($T = 15$ fs, $\Omega = 1.5$). (a), (b) local frequency shift, spectra, and longitudinal distortion of the pulse from WAKE simulations. The pulse propagates to the right. (a) the pulse at the plasma entrance ($z = 0$) and (b) at the point of electron dephasing ((b.1) $z = 2.03$ mm, (b.2) $z = 3.07$ mm). Grayscale is the absolute value of the Wigner transform (9) in arbitrary units; black curves are lineouts of the instantaneous frequency (in units of ω_0) extracted from the complex pulse envelope. Top inset: normalized intensity on axis. Right inset: radially integrated spectral power, $S(\omega) = \int_0^\infty \omega^2 |a(r,z,\omega)|^2 r dr$, in arbitrary units. (c) Electron spectra at dephasing from CALDER-Circ simulation. Black curves in (c.1) and (c.2) show the reference case spectra ($z = 2.03$ mm). Red (dark gray) in (c.2): Case B ($z = 3.07$ mm). Negative chirp of stack B doubles the energy of the QME signal, while suppressing the flux in the tail by more than an order of magnitude.

to 10 fs, according to **Figure 2(d)**) and then almost explosively elongates as newly generated mid-IR radiation slides into the bubble. Continuous injection ensues, building up the massive energy tail shown in **Figure 3(c.1)**, containing three-quarters of the charge accelerated above 50 MeV. CALDER-Circ simulation corroborates and adds more details to this unfavorable scenario. **Figure 4** links self-injection to the bubble evolution. The bubble size (shown in (a)) is defined as the length of the accelerating phase on axis (i.e. the length of the region inside the bubble where the longitudinal electric field is negative). Panel (b) tracks accumulation of the charge in the first two buckets, counting only electrons with $E > 50$ MeV. The collection phase space (longitudinal momenta of electrons at dephasing vs. their initial positions, (c)) and collection volume (initial positions of electrons with $E > 50$ MeV at dephasing, (e)) parameterize the energy gain of electrons with their initial coordinates. As the reference pulse adjusts for self-guiding, its spot size oscillates at least once. **Figure 4(a)** shows that it is during this early

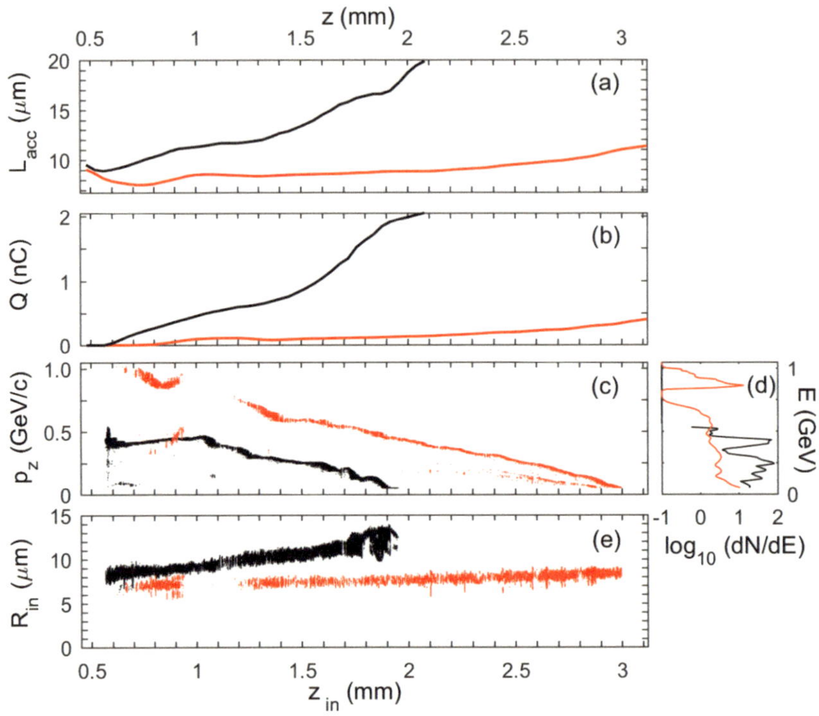

Figure 4. Stacking reduces expansion of the bubble, suppressing the dark current, avoiding the buildup of a low-energy tail in electron spectra (CALDER-Circ simulations). Black: the reference case. Red/dark gray: case B. (a) Length of the accelerating phase on axis vs. propagation length. (b) Charge accelerated in the first two buckets. (c) Longitudinal collection phase space of electrons from the first two buckets. (d) Energy spectra. These are identical to those in **Figure 3** (c.2), yet shown on a logarithmic scale, to evaluate suppression of the tail. (e) Collection volume. Data in plots (c)–(e) correspond to electron dephasing, $z = 2.03$ mm in the reference case and 3.07 mm in case B. The piecewise negative chirp of the stacked driver suppresses expansion of the bubble, reducing the flux in the energy tail by more than an order of magnitude, while doubling the energy of QME component.

stage (between $z \approx 0.55$ and 1.3 mm) that the bubble expands, injecting electrons, and then stabilizes, forming the QME bunch. Self-compression of the pulse starts early. As the optical shock builds up, the bubble starts to expand. Explosive expansion after $z \approx 1.3$ mm, with an almost 65% increase in size by the dephasing point ($z = 2.03$ mm), adds an extra 1.25 nC to the energy tail (in effect, multiplying the tail charge by a factor of 6). Beam loading saturates injection near dephasing, eventually destroying the bubble.

Figure 3(**a.2**)−(**c.2**) shows that the piecewise negative chirp turns the tide, enabling acceleration through dephasing without sacrificing e-beam quality. As a collateral benefit, electron energy doubles against the scaling-prescribed limit of the reference case. **Figure 4(c)** shows that, in case B, the injection starts later than in the reference case, while the bucket contracts rather than stabilizes around $z \approx 1.3$ mm, expelling one-third of the earlier injected charge. This reduces the bunch charge by a factor 8 in comparison with the reference case. Yet, from the data in **Table 1**, this reduction comes from clipping the bunch, from 5.5 fs to less than a femtosecond, with the average current preserved (≈ 88 kA). As soon as the QME bunch forms, the resilience of the stack to self-compression keeps it almost background-free (cf. **Figures 3(c.2)**

http://dx.doi.org/10.5772/intechopen.71679

Parameter (unit)	$\langle z \rangle$ (mm)	Q (pC)	$\langle E \rangle$ (MeV)	σ_E (MeV)	σ_τ (fs)	σ_α (mrad)	$\varepsilon_\perp^n$ (mm mrad)	$\langle I \rangle$ (kA)	B_n (A m^{-2})	W (mJ)
QME bunches										
Reference	2.027	493.5	426.5	25.7	5.5	2.93	0.69	89.7	0.38×10^{17}	210.4
Case A	2.148	288.8	524.8	26.3	3.8	2.75	0.64	76.2	0.38×10^{17}	151.5
Case B	1.473	73.8	442.9	31.8	0.85	2.16	0.3994	87.3	1.1×10^{17}	32.7
Case B, dephasing	3.067	73.8	881.9	28.6	0.85	1.35	0.3994	87.3	1.1×10^{17}	65.1
Energy tails ($E > 50$ MeV)										
Reference	2.017	1454	212.9	67.0	11.0	9.0	—	132	—	309.5
Case A	2.1425	363.3	253.2	113.4	5.5	6.52	—	66.1		91.2
Case B	1.4735	27.55	114.7	43.25	0.33	4.15	—	82.7	—	3.2
Case B, dephasing	3.062	329.8	298.4	174.6	8.84	5.16	—	37.3	—	98.4

Only particles from the first bucket are included. $\langle z \rangle$ is a longitudinal position of the beam centroid; Q is the charge; $\langle E \rangle$ is the mean energy; σ_E is the energy variance; σ_τ is the root-mean-square bunch length; σ_α is the root-mean-square divergence; $\varepsilon_\perp^n$ is the root-mean-square normalized transverse emittance; $\langle I \rangle = Q/\sigma_\tau$ is the average current; B_n is the 5-D brightness; W is the total energy of the bunch.

Table 1. Electron beam statistics.

and **4(d)**). In summary, stacking changes the system dynamics as follows. First, per **Figure 2(a)**, the pulse energy loss is merely one-quarter, in stark contrast with 60% of the reference case. Second, as is seen in **Figure 2(d)**, the stack reaches full compression at the point of electron dephasing rather than halfway through. **Figure 3(b.2)** shows that the energy of the stacked driver finally concentrates in a spike 2.5 optical cycles long, with the instantaneous frequency almost uniform along the pulse body. There is no sign of photon phase space rotation, with a mid-IR tail protruding into the bubble, nor there is a signature of a subcycle rising edge (the key feature of the reference scenario). Hence, the electron density pileup inside the compressed stack is minimal (cf. Figure 6 of Ref. [24]). Thereby, the resulting 7-fold reduction in the bubble expansion rate, evaluated from **Figure 4(a)**, reduces the average flux in the tail by a factor of 16 and the charge by a factor of 6. The QME peak dominates the electron spectrum, having the mean energy twice as high, and 5-D brightness a factor of 3.3 higher than its reference counterpart. Notably, the boost in energy has little to do with beam loading. As mentioned earlier, the current density in the QME bunches (the key factor defining the effect [26]) is almost the same in both cases. As expected, the WAKE test particle simulations show that the beam loading reduces electron energy by 25% in the reference case and merely by a few percent in case B. Hence, three-quarters of the observed energy boost are due to the favorable changes in the driver dynamics and quasistatic plasma response brought about by photon engineering.

Collection volume presented in **Figure 4(e)** indicates that only electrons with initial radial positions such as those that enter the bubble sheath are trapped and accelerated. There is no sign of transient injection from the near-axis region. The injection candidates fill a thin cylindrical shell with a radius slightly smaller than the bubble radius, accurately reproducing evolution of the pulse spot size in the cross-section at the highest intensity, which agrees with the matching condition (1). Indeed, for the 70 TW, 20 fs reference pulse, the power ratio $P/P_{cr} = 16.25$ yields the matched spot $r_m \approx 9.4$ μm. The collection radius in **Figure 4(e)** varies by ± 1 μm from this value through 70% of the acceleration distance. The dephasing length, calculated with this r_m, is $L_d \approx 1.65$ mm, which is within 10% of the value estimated from **Figure 4(e)** (black markers). The estimated energy gain (2) is 430 MeV, nearly the same as the simulated $\langle E \rangle \approx 426.5$ MeV. Thus, the QME bunch of the reference scenario complies with the scaling predictions exceptionally well. Yet, accumulation of the low-energy tail ruins the beam by the end of acceleration.

In case B, slow self-compression of the rigid head delays dephasing. Applying the scaling formulae to the 35 TW, 20 fs head with the wavelength $\lambda_{head} = (2/3)\lambda_0 \approx 0.533$ μm (so that $P/P_{cr} \approx 3.6$), we find $r_m \approx 7.3$ μm. Again, this value of r_m deviates from the collection radius shown in **Figure 4(e)** (red markers) by less than ± 1 μm throughout the entire interaction. The estimated dephasing length, $L_d \approx 2.95$ mm, is 15% longer than that obtained in the CALDER-Circ simulation. At the same time, the estimate of the energy gain at dephasing (using a generic formula (5) of Ref. [18]) is only 630 MeV, which is 30% lower than the gain obtained in the CALDER-Circ simulation with a bi-color stack. Even though, according to the scaling, using the head alone should boost electron energy effectively, the presence of the unshifted tail is important. As the rigid head of the stack plows through the plasma, driving the wake, the flapping of the slightly mismatched tail inside a soft channel (electron density bubble) controls the bubble radius, thus determining kinetics of self-injection [25]. The presence of the tail is thus essential to maintain sufficiently high charge (or modulate the beam current [25]). Since the tail rides inside the bubble devoid of electrons, it remains uncompressed. As the head starts experiencing red shift, the GVD-delayed red-shifted radiation superimposes onto the smooth profile of the tail. The combination of the two does not permit formation of a subcycle rising edge (compare **Figure 3(b.1)** and **3(b.2)**), which avoids uncontrollable expansion of the bubble.

Table 1 quantitatively assesses improvements in e-beam quality, showing parameters of QME bunches and energy tails at dephasing. Statistics of case B are complemented with the data taken at the point where the bunch energy matches the energy gain in the reference case ($E \approx 430$ MeV, $z \approx 1.47$ mm). **Table 1** presents the metrics that are essential to evaluate the 5-D brightness, $B_n = 2\langle I \rangle \left(\pi \varepsilon_\perp^n\right)^{-2}$, the quantity defining a capability of the beam to drive a TS-based γ-ray source [76]. Here, $\langle I \rangle = Q/\sigma_\tau$ is the mean current; Q is the bunch charge; σ_τ is the root-mean-square bunch length; and $\varepsilon_\perp^n = 2^{-1/2}\left(\left(\varepsilon_x^n\right)^2 + \left(\varepsilon_y^n\right)^2\right)^{1/2}$, where $\varepsilon_i^n = (m_e c)^{-1}\left(\left(\langle p_i^2 \rangle - \langle p_i \rangle^2\right)\left(\langle r_i^2 \rangle - \langle r_i \rangle^2\right) - \left(\langle p_i r_i \rangle - \langle p_i \rangle \langle r_i \rangle\right)^2\right)^{1/2}$ is the root-mean-square normalized transverse emittance. Statistics for case B show that the QME bunch progresses through dephasing with the normalized transverse emittance conserved, as should be the case for the adiabatically slowly varying structure. Simulating e-beam dynamics, while conserving the emittance better than in the fourth digit, became possible due to elimination of the numerical Cherenkov radiation [86] in CALDER-Circ. Any degradation of the Thomson γ-ray signal observed in the simulations must be thus attributed to the physical causes rather than to numerical artifacts. The QME bunch B,

http://dx.doi.org/10.5772/intechopen.71679

apart from 3.2% energy spread, sub-fs duration, and 400 nm emittance (about half of that of the reference case), has the 5-D brightness 1.1×10^{17} A/m^2, preserved throughout acceleration. This is most encouraging for using the beam as a driver of a high-flux Thomson source [76].

4. Stack-driven electron beams generate high-flux, femtosecond γ-ray pulses via Thomson scattering

The capability of a stack-driven LPA to suppress the low-energy background and to increase electron energy gain, while preserving 100-pC scale charge and 100-kA average current, boosting the e-beam brightness beyond 10^{17} A/m^2, is an asset for the design of radiation sources.

4.1. Improving performance of the γ-ray source: suppressing emission of low-energy photons and boosting the energy of quasi-monochromatic signal

A broadband e-beam accelerated with a transform-limited pulse is poorly suited to produce quasi-monochromatic γ-ray pulse via the TS mechanism. **Figure 5(a)** demonstrates the phase space of the e-beam at dephasing, with a QME component accompanied with a massive energy tail. We separate the macroparticles making up the QME e-bunch and the tail, as shown in **Figure 5(a)**, and carry out two sets of TS simulations with these initial conditions. The partial photon spectra displayed in **Figure 5(c)** reveal modest energy of γ-photons emitted by the

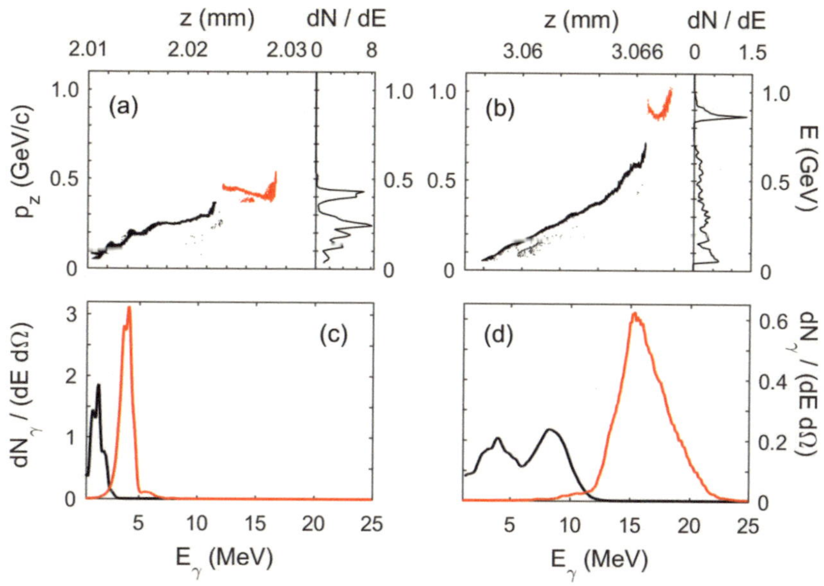

Figure 5. The stack-driven LPA delivers a low-background QME electron bunch. Thomson scattering from this bunch produces a quasi-monochromatic, sub-fs γ-ray pulse. Electron beams are extracted from CALDER-Circ simulations at dephasing. (a), (c) reference case. (b), (d) case B. (a), (b) longitudinal phase space of the bunch; inset: energy spectrum (in units MeV^{-1}). (c), (d) γ-ray flux in the direction of e-beam propagation (in units 10^{12} MeV^{-1} sr^{-1}). QME components of e-beams and corresponding quasi-monochromatic components of γ-ray pulses are depicted in red (dark gray).

QME bunch in the direction of its propagation, $\langle E_\gamma \rangle \approx 3.85$ MeV. This is too low to meet the needs of nondestructive inspection systems for special nuclear materials; these require photon energy tunable in the range 5−15 MeV [49]. **Figure 5(c)** reveals a rather high ratio of noise to the quasi-monochromatic signal, 1:2, on average. Conversely, the virtual lack of the low-energy background in case B favorably reflects on the TS signal. The latter, from **Figure 5(d)**, has an average noise-to-signal ratio twice as low, 1:4, and nearly a factor 4 higher mean photon energy, $\langle E_\gamma \rangle \approx 16$ MeV. Yet, the signal bandwidth is rather high, 15.7%.

From a single-particle theory [50], both variation in energy of an electron and a misalignment of its trajectory with the propagation axis of the ILP tend to reduce the photon energy, thus contributing to the photon energy spread. The QME bunch of case B, apart from having a 3.2% energy spread, has a rather high root-mean-square divergence, $\sigma_\alpha = 1.35$ mrad, or $2.33\langle \gamma_e \rangle^{-1}$. Here, $\langle \gamma_e \rangle$ is the mean Lorentz factor of the bunch, and $\sigma_\alpha = 2^{-1/2}\left(\sigma_x^2(\alpha) + \sigma_y^2(\alpha)\right)^{1/2}$, where $\sigma_i(\alpha) = \langle p_z \rangle^{-1}\left(\langle p_i^2 \rangle - \langle p_i \rangle^2\right)^{1/2}$.

TS simulations with the reduced phase space of e-beam help identify the primary contributor to the photon energy spread. First, we plot in **Figure 6** (black in both panels) the spectrum of γ-photons emitted by the bunch with a complete 6-D phase space (the region of phase space depicted with red/dark gray markers in **Figure 5(b)**). The signal with a 15.7% energy spread is centered at $\langle E_\gamma \rangle = 16$ MeV. Then, transverse momenta of all macroparticles are set to zero, while their longitudinal momenta, p_z, are unchanged. This preserves the energy spread while zeroing out the divergence. Lastly, $p_z = \langle p_z \rangle = 1725.8 m_e c$ is assigned to all electrons, while p_x and p_y are unchanged. This preserves mrad-scale divergence of the bunch, while nearly zeroing out the energy spread. In the zero-divergence case (spectrum depicted in red/dark gray in **Figure 6(a)**), the mean photon energy increases to 18 MeV, while the energy spread stays at 12.8%. In stark contrast, the case with a near-zero energy spread yields a TS signal with a *subpercent* energy spread, centered at $E_\gamma \approx 4\langle \gamma_e \rangle^2 E_{\text{int}} \approx 18.4$ MeV (red/dark gray in **Figure 6 (b)**). Thus, the γ-ray signal receives its large (more than 10%) bandwidth almost entirely from a few-percent electron energy spread. Further steps in optimization of the LPA should aim to

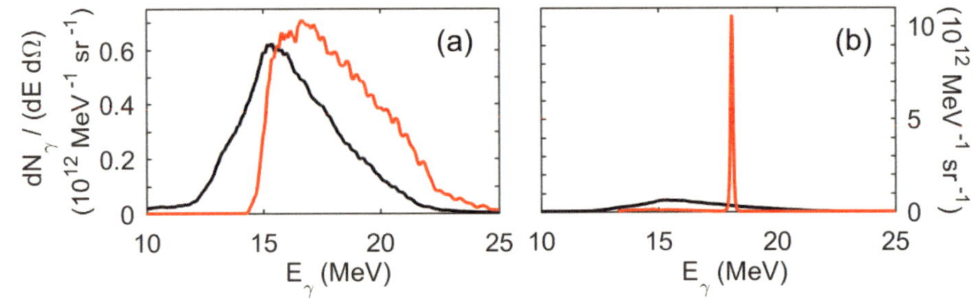

Figure 6. Energy spread in the electron bunch determines the energy spread of emitted γ-photons. Black: TS spectrum from the simulation using complete phase space of electron bunch (same as red (dark gray) in **Figure 5(c)** and **5(d)**). Red (dark gray): simulations with the reduced phase space of electrons, with (a) zero dispersion of transverse momentum (zero divergence) and (b) zero dispersion of longitudinal momentum (viz. almost vanishing energy spread).

reduce the energy spread below 1%. One practical way to do it (at the expense of reduction in charge and, hence, brightness) is to select electrons from narrow energy intervals within the e-beam bandwidth by selective focusing with highly chromatic magnetic quadrupole lenses [90] before the collision of e-beam with the ILP. (At the same time, the magnetic quadrupole will disperse the residual energy tail.)

The collimation of high-energy γ-photons and the number of photons in the observation cone are important metrics for applications. To evaluate the reduction in photon energy and flux with an increase in the observation angle (viz. to estimate the effective apex angle of the photon emission cone), we select the macroparticles making up the QME e-bunch and carry out the TS simulation with these initial conditions. We detect the photons scattered in and out of the ILS polarization plane, in the direction of e-beam propagation ($\theta = 0$) and at small angles measured from the direction of e-beam propagation, up to the root-mean-square divergence angle, $\theta = \sigma_\alpha = 2.33\langle\gamma_e\rangle^{-1}$. As the scattering angle increases to $\theta = \langle\gamma_e\rangle^{-1}$, the mean photon energy drops by 25%, the energy spread staying at the 15–20% level. At the same time, the photon flux drops 10-fold. To a good approximation, there are virtually no photons with the energies above 10 MeV outside the observation cone of apex angle $2\theta = 2\langle\gamma_e\rangle^{-1}$. Thus, to estimate the number of QME high-energy photons scattered in the direction of e-beam propagation, we choose conservatively the observation solid angle $\Delta\Omega_\gamma = (\pi/2)\langle\gamma_e\rangle^{-2}$, that is, the solid angle of the cone with an apex angle $2\theta = \sqrt{2}\langle\gamma_e\rangle^{-1}$ ($\approx 2 \times 0.69$ (2×0.41) mrad for the beam A(B) at dephasing). We take the photon flux corresponding to the direct backscattering ($\theta = 0$), integrate it over the energy, and multiply the result by $\Delta\Omega_\gamma$.

Table 2 presents statistics of quasi-monochromatic γ-ray signals corresponding to the QME entries in **Table 1**. The entries corresponding to the beam B show the photon yield over 1.5×10^6 per pulse. This is comparable to the experimental yields with 100-MeV scale e-beams, 3×10^5 to 10^7 [66, 67, 70], calculated for the entire forward hemisphere, and integrated over the entire broad bandwidth of the γ-ray beam. Yet, our highest-energy photons reach 16 MeV while preserving a 16% energy spread and microsteradian collimation, which is strikingly better than 50–100% spread and millisteradian collimation reported for the sub-MeV photons [66, 67, 70].

Parameter	$\langle E_\gamma\rangle$ (MeV)	σ_E (MeV)	$\Delta\Omega_\gamma$ (μsr)	N_γ (10^6)	W_γ (μJ)
Reference	3.85	0.72	2.25	8.95	5.5
Case A	5.67	0.97	1.49	5.08	4.6
Case B	4.36	0.93	2.09	1.52	1.1
Case B, dephasing	16.0	2.51	0.53	1.58	4.0

Corresponding energy spectra are depicted in **Figure 5(c)** (reference); **Figure 7(c)** (case A); **Figure 7(d)**, red (case B before dephasing); and **Figure 7(d)**, black (Case B at dephasing). $\langle E_\gamma\rangle$ is the mean energy; σ_E is the energy variance; N_γ and $W_\gamma = N_\gamma\langle E_\gamma\rangle$ are the number of photons and energy radiated into the observation solid angle $\Delta\Omega_\gamma = (\pi/2)\langle\gamma_e\rangle^{-2}$ in the direction of e-beam propagation.

Table 2. Statistics of γ-rays emitted by the QME bunches with parameters from **Table 1**.

4.2. Tuning the energy and flux of the quasi-monochromatic γ-ray signal

The structure of e-beam phase space is sensitive to variations in the drive pulse evolution brought about by changes in the initial conditions. This responsiveness improves parameters of the beam to make it suitable to drive a quasi-monochromatic radiation source based on Thomson scattering, a process demanding exceptional beam quality. **Figure 7** displays the capability of the all-optical Thomson source to produce γ-ray pulses with different characteristics (such as a mean energy and flux) while maintaining high yield and low background. This flexibility is demonstrated without changing the energy of the stack or the frequency ratio between its components. This is an asset to applications, which often need to adjust to new conditions in a timely fashion, avoiding a major upgrade of the laboratory.

Changing the e-beam energy, while keeping the brightness fixed, is one option offered by case B. Standard targets, such as printed gas cells of variable length [81], permit the necessary reductions in the plasma length. **Figure 7(b)** shows progress of the e-beam through dephasing, from the energy matching the maximal gain in the reference case (≈430 MeV) to 882 MeV. Notably, the QME bunch reaches dephasing, while maintaining $B_n = 1.1 \times 10^{17}$ A/m^2. This conserves the number of photons emitted into the μsr-scale detection angle, $N_\gamma \approx 1.55 \times 10^6$. Thus, a quasi-monochromatic (15.7–21.3% energy variance) 0.85 fs γ-ray pulse with the mean energy tunable between 4 and 16 MeV (1.3–4.7 GW average power) can be produced.

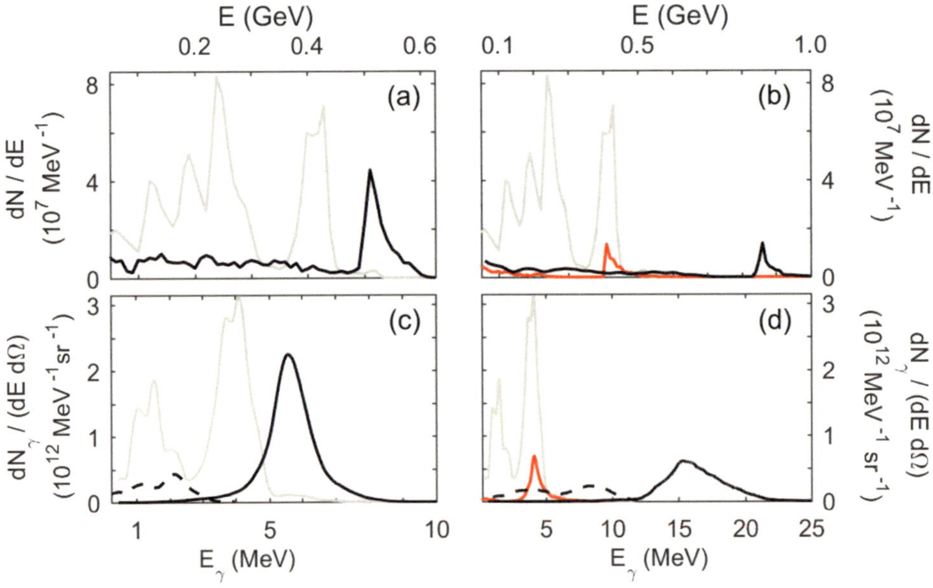

Figure 7. The stack with fully overlapped components (case A, panels (a), (c)) preserves high electron and γ-ray flux similar to that of the reference case, while suppressing the background. The stack with a blue component advanced in time (case B, panels (b), (d)) permits doubling electron energy against the reference case, increasing the photon energy by a factor 3 in comparison to case A, at the expense of reduction in flux. (a), (b) electron energy spectra. (c), (d) spectra of γ-photons emitted in the direction of e-beam propagation. Gray: Spectra of the reference case. Black: electron and photon spectra corresponding to e-beams extracted at dephasing ($z \approx 2.15$ mm, case A; $z \approx 3.07$ mm, case B). Dashed curves in (c) and (d): spectra of photons emitted by the electron energy tails. Red (dark gray): E-beam of case B, extracted prior to dephasing, $z \approx 1.47$ mm.

Reducing the time delay between stack components increases photon yield at relatively low energies ($\langle E_\gamma \rangle \leq 10$ MeV), the ultimate example of which is case A (full overlap). The physical difference between cases A and B is remarkable. Stack A is an incoherent mix in the fashion of Ref. [80] rather than a pulse with a negative piecewise chirp. As the stack A components plow through the plasma, they both ride on the down-slope of the nonlinear index, such as depicted, e.g. in **Figure 7** of Ref. [16] or **Figure 5** of Ref. [22]. From **Figure 1(b.2)**, the frequency unshifted component, E_0, red-shifts and compresses to nearly a single cycle, in the same fashion as the reference pulse does (cf. **Figure 1(a.2)**). At the same time, the blue-shifted E_{head} remains virtually intact. **Figure 1(b.2)** also shows that the stack components stay together, accumulating merely a two-cycle delay due to the difference in their group velocities. The stack thus does not break up longitudinally. The superposition of compressed E_0 and almost intact E_{head} does not allow for the formation of a subcycle rising edge (optical shock) such as develops in the reference case (**Figure 1(a.2)**). The undelayed E_{head} is thus akin to a rigid exoskeleton protecting the vulnerable E_0.

Deformation of a stack with fully overlapped components, the process that defines the dephasing length [18], is dominated by rapid self-compression of the least resilient stack component, E_0. As E_0 self-compresses (and hence, the stack self-compresses) at the same rate as the drive pulse in the reference scenario, the dephasing length remains almost unchanged, and the boost in electron energy (by 23%) is unremarkable. **Table 1** shows that the QME bunch of case A is not very different from its reference counterpart (note that their 5-D brightness is the same.) Conversely, the electron energy tail in case A is suppressed, with a factor 4 reduction in charge and 7 in average flux. This drastically improves the noise-to-signal ratio of emitted γ-rays, from roughly 1:2 in the reference case to better than 1:5 in case A, as may be evaluated from photon spectra in **Figure 7(c)**. **Table 2** shows that the γ-ray pulse, in case A, contains about 5×10^6 photons, a factor 3.3 higher yield than in case B. Further, these photons receive 50% boost in energy with respect to the reference case[2]. Changing the time delay between the stack components is thus a proper way to control the TS photon yield and energy.

4.3. Final notes on all-optical control of quasi-monochromatic Thomson sources

The results presented above offer an opportunity for tuning the γ-ray energy and flux in a broad range, without changing the total (Joule-per-pulse) laser energy in the LPA.

Particularly, the interval of photon energy variation, 4–16 MeV, with the yield $(1.5 - 5) \times 10^6$ photons per pulse, meets the needs of nuclear photonics applications [49]. Raising the photon yield per second is technologically possible. Using a longer (up to 2.5 ps) ILP with the same spot size and amplitude ($a_{\text{int}} = 0.1$) may increase the yield by up to an order of magnitude. This long ILP is still shorter than its half-Rayleigh length. This is sufficient to preserve the nearly plane-wave character of the interaction with the e-bunch. As the energy (0.25 J) of the long ILP is still below the stack energy (1.4 J), their repetition rates may be matched. Then, using a kW-scale laser amplifier [77] permits the production of Joule scale pulses (LPA stack

[2] Increasing the ILP frequency may substantially increase the photon energy, while preserving the high flux [72].

and ILP) at a few hundred Hz, increasing the photon yield beyond 10^9 ph/s. Apart from keeping the photon yield at a competitive level, this repetition rate permits real time optimization of the experiment [82, 83], not possible with one shot per hour PW facilities [73].

The estimated 10^9 ph/s yield is sufficiently high to make the tunable TS source interesting for the design of a nondestructive inspection system for special nuclear materials. Simulations, based on data from recent detection experiment [61], indicate that the TS γ-ray flux of 10^6 ph/s, with a 5% signal bandwidth and a 10 Hz repetition rate, is sufficient to identify a nuclear resonance fluorescence peak from a 1 kg of highly enriched uranium within 10 minutes. The higher flux accessible with our source would compensate for its still significant (up to 20%) bandwidth. Alternatively, frequency chirping of the ILP may reduce the photon energy spread [51]. Given the unconventional U-shape of electron momentum chirp in the QME e-bunch (cf. **Figure 5(a)** and **5(b)**), this topic deserves special consideration but is beyond the scope of this chapter.

Lastly, e-beam optimization studies, using stacks with different frequency ratios [24], show that, as long as the time delay and energy partition between the stack components are fixed, and $\Omega \geq 1.25$, the electron energy gain is quite insensitive to the frequency ratio. Even though the particle flux dN/dE in the QME bunch drops as $\Omega \to 2$, reducing the quasi-monochromatic γ-ray yield, the TS signal still has a quality far exceeding that accessible in the reference scenario. This observation permits a considerable technological flexibility in a practical realization of this concept. Frequency shifting on the modest scale ($\Omega \leq 1.5$) can be accomplished with a Raman cell, with subsequent conventional chirped-pulse amplification [91–93]. Alternatively, energy-efficient methods of frequency doubling may be applied.

5. Summary and outlook

In a conventional LPA, a cavity of electron density maintained by the radiation pressure of a single narrow-bandwidth laser pulse accelerates electrons self-injected from the ambient plasma. Deformations of the bucket, which carry on in lock-step with the deformations of the optical driver, determine the structure of the e-beam phase space. Optimizing the nonlinear evolution of the drive pulse is an essential element of LPA design, offering new avenues to coherently control e-beam phase space on the femtosecond scale.

Compact sources of quasi-monochromatic γ-photons, based on the TS mechanism, are highly sensitive to the quality and phase space structure of the driving GeV-scale e-beams. Reaching sufficient e-beam brightness and energy, while maintaining a modest facility footprint and high repetition rate, is a major challenge for a traditional LPA. The first roadblock is the limit on electron energy imposed by dephasing, with unavoidable beam contamination with a low-energy background, while the second is the low repetition rate of petawatt-scale lasers (which limits the dosage, frustrating applications). Reducing the energy in the drive pulse to a sub-Joule level may alleviate the latter, yet aggravating the former. Our simulations show the way to resolve this conflict, by synthesizing the LPA drive pulse by incoherently stacking collinearly propagating 10-TW-scale pulses of different wavelengths. Stacking introduces a frequency bandwidth sufficient to compensate the red shift imparted by the wake excitation.

Unlike a single, transform-limited pulse, the stack is nearly immune to degradation while driving the bubble in a dense plasma ($n_0 \sim 10^{19}$ cm^{-3}). Advancing the blue-shifted component of the stack in time emulates the negative frequency chirp [16, 23–25]. This delays dephasing of the electrons, doubling their energy compared to the scaling predictions, using no manipulations of a few mm-length gas target. Importantly, immunity of the stacked driver to self-compression keeps the low-energy electron flux so modest as to almost avoid contamination of TS γ-ray pulse with low-energy photons.

Simulation data presented here show that increasing the delay between the stack components, while keeping the same total laser energy and frequency ratio, permits increasing electron energy from 525 to 900 MeV, boosting the 5-D brightness nearly three-fold. This puts generation of 10 MeV-scale, few-GW, quasi-monochromatic, femtosecond-length γ-ray pulses via Thomson scattering within reach of existing laser technology. Energy of these pulses, containing up to 5×10^6 photons into the microsteradian cone, may be tuned in the range 4–16 MeV and, possibly, beyond, while the low-energy photon background remains insignificant. Increasing the LPA repetition rate to kHz level, at affordable average power, promises boosting photon yield beyond 10^9 ph/s, making tunable, all-optical TS γ-ray sources interesting for applications [48, 49, 61]. As a further development, focusing the stack components differently [52], or propagating the stack in a channel [24, 25], enables generation of a train of GeV-scale, ultrabright electron bunches with a femtosecond synchronization. These unconventional comb-like e-beams may emit polychromatic trains of high-flux γ-ray pulses consisting of a few distinct energy bands, in the range 3–17 MeV [52]. The natural mutual synchronization of fs-length e-bunches and γ-ray pulses may be an asset to nuclear pump-probe experiments. With a γ-ray beam spectrally resolved, each beamlet may give a "movie frame" on a femtosecond time scale to image ultrafast phenomena in a dense matter [74].

Acknowledgements

The US DOE Grant DE-SC0008382 and National Science Foundation Grants PHY-1104683 and PHY-1535678 have supported the work of SYK and BAS. Thomson scattering simulations were completed utilizing high-performance computing resources of the Holland Computing Centre (HCC) of the University of Nebraska.

Author details

Serge Y. Kalmykov[1]*, Xavier Davoine[2], Isaac Ghebregziabher[3] and Bradley A. Shadwick[1]

*Address all correspondence to: s.kalmykov.2013@ieee.org

1 Department of Physics and Astronomy, University of Nebraska–Lincoln, Lincoln, NE, USA

2 CEA DAM DIF, Arpajon, France

3 The Pennsylvania State University, Hazleton, PA, USA

References

[1] Budker GJ. Relativistic stabilized electron beam: I. Physical principles and theory. In: Regenstreif E, editor. Proceedings of CERN Symposium on High Energy Accelerators and Pion Physics; 11-23 June 1956; Geneva, Switzerland. Geneva, Switzerland: CERN; 1956. p. 68-75. DOI: 10.5170/CERN-1956-025.68

[2] Veksler VI. Coherent principle of acceleration of charged particles. In: Regenstreif E, editor. Proceedings of CERN Symposium on High Energy Accelerators and Pion Physics; 11-23 June 1956; Geneva, Switzerland. Geneva, Switzerland: CERN; 1956. p. 81-83. DOI: 10.5170/CERN-1956-025.80

[3] Fainberg IaB. The use of plasma waveguides as accelerating structures in linear accelerators. In: Regenstreif E, editor. Proceedings of CERN Symposium on High Energy Accelerators and Pion Physics; 11-23 June 1956; Geneva, Switzerland. Geneva, Switzerland: CERN; 1956. p. 84-90. DOI: 10.5170/CERN-1956-025.84

[4] Mangles SPD, Murphy CD, Najmudin Z, Thomas AGR, Collier JL, Dangor AE, et al. Monoenergetic beams of relativistic electrons from intense laser–plasma interactions. Nature (London). 2004;**431**:535-538. DOI: 10.1038/nature02939

[5] Geddes CGR, Toth Cs, van Tilborg J, Esarey E, Schroeder CB, Bruhwiler D, et al. High-quality electron beams from a laser wakefield accelerator using plasma-channel guiding. Nature (London). 2004;**431**:538-541. DOI: 10.1038/nature02900

[6] Faure J, Glinec Y, Pukhov A, Kiselev S, Gordienko S, Lefebvre E, et al. A laser–plasma accelerator producing monoenergetic electron beams. Nature (London). 2004;**431**:541-544. DOI: 10.1038/nature02963

[7] Blumenfeld I, Clayton CE, Decker F-J, Hogan MJ, Huang C, Ischebeck R, et al. Energy doubling of 42 GeV electrons in a metre-scale plasma wakefield accelerator. Nature (London). 2007;**445**:741-744. DOI: 10.1038/nature05538

[8] Corde S, Adli E, Allen JM, An W, Clarke CI, Clayton CE, et al. Multi-gigaelectronvolt acceleration of positrons in a self-loaded plasma wakefield. Nature (London). 2015;**524**: 442-445. DOI: 10.1038/nature14890

[9] Tajima T, Dawson JM. Laser electron accelerator. Physical Review Letters. 1979;**43**(4):267-270. DOI: 10.1103/PhysRevLett.43.267

[10] Kalmykov SY, Gorbunov LM, Mora P, Shvets G. Injection, trapping, and acceleration of electrons in a three-dimensional nonlinear laser wakefield. Physics of Plasmas. 2006;**13**(11): 113102. DOI: 10.1063/1.2363172

[11] Matlis NH, Reed SA, Bulanov SS, Chvykov V, Kalintchenko G, Matsuoka T, et al. Snapshots of laser wakefields. Nature Physics. 2006;**2**:749-753. DOI: 10.1038/nphys442

[12] Rosenzweig JB, Breizman B, Katsouleas T, Su JJ. Acceleration and focusing of electrons in two-dimensional nonlinear plasma wake fields. Physical Review A. 1991;**44**(10):R6189-R6192. DOI: 10.1103/PhysRevA.44.R6189

[13] Mora P, Antonsen TM Jr. Electron cavitation and acceleration in the wake of ultraintense, self-focused laser pulse. Physical Review E. 1996;**53**(3):R2068-R2071. DOI: 10.1103/PhysRevE.53.R2068

[14] Pukhov A, Meyer-ter-Vehn J. Laser wake field acceleration: The highly non-linear broken-wave regime. Applied Physics B: Lasers & Optics. 2002;**74**(4–5):355-361. DOI: 10.1007/s003400200795

[15] Kalmykov SY, Yi SA, Beck A, Lifschitz AF, Davoine X, Lefebvre E, et al. Numerical modelling of a 10-cm-long multi-GeV laser wakefield accelerator driven by a self-guided petawatt pulse. New Journal of Physics. 2010;**12**(4):045019. DOI: 10.1088/1367-2630/12/4/045019

[16] Kalmykov SY, Shadwick BA, Beck A, Lefebvre E. Physics of quasi-monoenergetic laser-plasma acceleration of electrons in the blowout regime. In: Andreev AV, editor. Femtosecond-Scale Optics. Rijeka, Croatia: InTech; 2011. pp. 113-138. DOI: 10.5772/25062

[17] Malka V. Laser plasma accelerators. Physics of Plasmas. 2012;**19**(5):055501. DOI: 10.1063/1.3695389

[18] Lu W, Tzoufras M, Joshi C, Tsung FS, Mori WB, Vieira J, et al. Generating multi-GeV electron bunches using single stage laser wakefield acceleration in a 3D nonlinear regime. Physical Review Accelerators and Beams. 2007;**10**(6):061301. DOI: 10.1103/PhysRevSTAB.10.061301

[19] Dong P, Reed SA, Yi SA, Kalmykov S, Shvets G, Downer MC, et al. Formation of optical bullets in driven plasma bubble accelerators. Physical Review Letters. 2010;**104**(13):134801. DOI: 10.1103/PhysRevLett.104.134801

[20] Li Z, Tsai H-E, Zhang X, Pai C-H, Chang Y-Y, Zgadzaj R, et al. Single-shot visualization of evolving laser wakefields using an all-optical streak camera. Physical Review Letters. 2014;**113**(8):085001. DOI: 10.1103/PhysRevLett.113.085001

[21] Kalmykov SY, Beck A, Yi SA, Khudik V, Shadwick BA, Lefebvre E, Downer MC. *electron* Self-injection into an evolving plasma bubble: The way to a dark current free GeV-scale laser accelerator. AIP Conference Proceedings. 2010;**1299**:174-179. DOI: 10.1063/1.3520309

[22] Kalmykov SY, Beck A, Yi SA, Khudik VN, Downer MC, Lefebvre E, et al. Electron self-injection into an evolving plasma bubble: Quasi-monoenergetic laser-plasma acceleration in the blowout regime. Physics of Plasmas. 2011;**18**(5):056704. DOI: 10.1063/1.3566062

[23] Kalmykov SY, Beck A, Davoine X, E Lefebvre E, Shadwick BA. Laser plasma acceleration with a negatively chirped pulse: All-optical control over dark current in the blowout regime. New Journal of Physics. 2012;**14**(3):033025. DOI: 10.1088/1367-2630/14/3/033025

[24] Kalmykov SY, Davoine X, Lehe R, Lifschitz AF, Shadwick BA. Optical control of electron phase space in plasma accelerators with incoherently stacked laser pulses. Physics of Plasmas. 2015;**22**(5):056701. DOI: 10.1063/1.4920962

[25] Kalmykov SY, Davoine X, Ghebregziabher I, Lehe R, Lifschitz AF, Shadwick BA. Controlled generation of comb-like electron beams in plasma channels for polychromatic Thomson γ-ray sources. Plasma Physics and Controlled Fusion. 2016;**58**(3):034006. DOI: 10.1088/0741-3335/58/3/034006

[26] Tzoufras M, Lu W, Tsung FS, Huang C, Mori WB, Katsouleas T, et al. Beam loading by electrons in nonlinear plasma wakes. Physics of Plasmas. 2009;**16**(5):056705. DOI: 10.1063/1.3118628

[27] Sun G-Z, Ott E, Lee YC, Guzdar P. Self-focusing of short intense pulses in plasmas. The Physics of Fluids. 1987;**30**(2):526-532. DOI: 10.1063/1.866349

[28] Mora P, Antonsen Jr TM. Kinetic modeling of intense, short laser pulses propagating in tenuous plasmas. Physics of Plasmas. 1997;**4**(1):217-229. DOI: 10.1063/1.872134

[29] Thomas AGR, Najmudin Z, Mangles SPD, Murphy CD, Dangor AE, Kamperidis C, et al. The effect of laser focusing conditions on propagation and monoenergetic electron production in laser wakefield accelerators. Physical Review Letters. 2007;**98**(9):095004. DOI: 10.1103/PhysRevLett.98.095004

[30] Thomas AGR, Mangles SPD, Murphy CD, Dangor AE, Foster PS, Gallacher JG, et al. Ultrashort pulse filamentation and monoenergetic electron beam production in LWFAs. Plasma Physics and Controlled Fusion. 2009;**51**(2):024010. DOI: 10.1088/0741-3335/51/2/024010

[31] Pai C-H, Chang Y-Y, Ha L-C, Xie Z-H, Lin M-W, Lin J-M, et al. Generation of intense ultrashort midinfrared pulses by laser-plasma interaction in the bubble regime. Physical Review A. 2010;**82**(6):063804. DOI: 10.1103/PhysRevA.82.063804

[32] Zhu W, Palastro JP, Antonsen Jr TM. Pulsed mid-infrared radiation from spectral broadening in laser wakefield simulations. Physics of Plasmas. 2013;**20**(7):073103. DOI: 10.1063/1.4813245

[33] Gordon DF, Hafizi B, Hubbard RF, Peano JR, Sprangle P, Ting A. Asymmetric self-phase nodulation and compression of short laser pulses. Physical Review Letters. 2003;**90**(21):215001. DOI: 10.1103/PhysRevLett.90.215001

[34] Beck A, Kalmykov SY, Davoine X, Lifschitz A, Shadwick BA, Malka V, Specka A. Physical processes at work in sub-30fs, PW laser pulse-driven plasma accelerators: Towards GeV electron acceleration experiments at CILEX facility. Nuclear Instruments and Methods in Physics Research Section A: Accelerators, Spectrometers, Detectors and Associated Equipment. 2014;**740**:67-73. DOI: 10.1016/j.nima.2013.11.003

[35] Banerjee S, Powers ND, Ramanathan V, Ghebregziabher I, Brown KJ, Maharjan CM, et al. Generation of tunable, 100–800 MeV quasi-monoenergetic electron beams from a laser-wakefield accelerator in the blowout regime. Physics of Plasmas. 2012;**19**(5):056703. DOI: 10.1063/1.4718711

[36] Bayramian AJ, Armstrong JP, Beer G, Campbell R, Cross R, Erlandson A, et al. High average power petawatt laser pumped by the mercury laser for fusion materials engineering. Fusion Science and Technology. 2009;**56**:295-300

[37] Fattahi H, Barros HG, Gorjan M, Nubbemeyer T, Alsaif B, Teysset CY, et al. Third-generation femtosecond technology. Optica. 2014;**1**(1):45-63. DOI: 10.1364/OPTICA.1.000045

[38] Cowan BM, Kalmykov SY, Beck A, Davoine X, Bunkers K, Lifschitz AF, et al. Computationally efficient methods for modelling laser wakefield acceleration in the blowout regime. Journal of Plasma Physics. 2012;**78**(4):469-482. DOI: 10.1017/S0022377812000517

[39] Kneip S, Nagel SR, Martins SF, Mangles SPD, Bellei C, Chekhlov O, et al. Near-GeV acceleration of electrons by a nonlinear plasma wave driven by a self-guided laser pulse. Physical Review Letters. 2009;**103**(3):035002. DOI: 10.1103/PhysRevLett.103.035002

[40] Kneip S, Nagel SR, Bellei C, Cheklov O, Clarke RJ, Delerue N, et al. Study of near-GeV acceleration of electrons in a non-linear plasma wave driven by a self-guided laser pulse. Plasma Physics and Controlled Fusion. 2011;**53**(1):014008. DOI: 10.1088/0741-3335/53/1/014008

[41] Froula DH, Clayton CE, Döppner T, Marsh KA, Barty CPJ, Divol L, et al. Measurements of the critical power for self-injection of electrons in a laser wakefield accelerator. Physical Review Letters. 2009;**103**(21):215006. DOI: 10.1103/PhysRevLett.103.215006

[42] Ralph JE, Clayton CE, Albert F, Pollock BB, Martins SF, Pak AE, et al. Laser wakefield acceleration at reduced density in the self-guided regime. Physics of Plasmas. 2010;**17**(5):056709. DOI: 10.1063/1.3323083

[43] Corde S, Ta Phuoc K, Lambert G, Fitour R, Malka V, Rousse A. Femtosecond x rays from laser-plasma accelerators. Reviews of Modern Physics. 2013;**85**(1):1-48. DOI: 10.1103/RevModPhys.85.1

[44] Lau YY, Fei He F, Umstadter DP, Kowalczyk R. Nonlinear Thomson scattering: A tutorial. Physics of Plasmas. 2003;**10**(3):2155-2162. DOI: 10.1063/1.1565115

[45] Sarachik ES, Schappert GT. Classical theory of the scattering of intense laser radiation by free electrons. Physical Review D. 1970;**1**(10):2738-2753. DOI: 10.1103/PhysRevD.1.2738

[46] Esarey E, Ride SK, Sprangle P. Nonlinear Thomson scattering of intense laser pulses from beams and plasmas. Physical Review E. 1993;**48**(4):3003-3021. DOI: 10.1103/PhysRevE.48.3003

[47] Ride SK, Esarey E, Baine M. Thomson scattering of intense lasers from electron beams at arbitrary interaction angles. Physical Review E. 1995;**52**(5):5425-5442. DOI: 10.1103/PhysRevE.52.5425

[48] Albert F, Anderson SG, Gibson DJ, Marsh RA, Wu SS, Siders CW, et al. Design of narrow-band Compton scattering sources for nuclear resonance fluorescence. Physical Review Accelerators and Beams. 2011;**14**(5):050703. DOI: 10.1103/PhysRevSTAB.14.050703

[49] Rykovanov SG, Geddes CGR, Vay J-L, Schroeder CB, Esarey E, Leemans WP. Quasi-monoenergetic femtosecond photon sources from Thomson scattering using laser plasma accelerators and plasma channels. Journal of Applied Physics B: Atomic, Molecular and Optical Physics. 2014;**47**(23):234013. DOI: 10.1088/0953-4075/47/23/234013

[50] Tomassini P, Giulietti A, Giulietti D, Gizzi LA. Thomson backscattering X-rays from ultra-relativistic electron bunchesand temporally shaped laser pulses. Applied Physics B: Lasers & Optics. 2005;**80**(4–5):419-436. DOI: 10.1007/s00340-005-1757-x

[51] Ghebregziabher I, Shadwick BA, Umstadter DP. Spectral bandwidth reduction of Thomson scattered light by pulse chirping. Physical Review Accelerators and Beams. 2013; **16**(3):030705. DOI: 10.1103/PhysRevSTAB.16.030705

[52] Kalmykov SY, Davoine X, Ghebregziabher I, Shadwick BA. Multi-color γ-rays from comb-like electron beams driven by incoherent stacks of laser pulses. AIP Conference Proceedings. 2017;**1812**:100001. DOI: 10.1063/1.4975899

[53] Hsu IC, Chu C-C, Yu C-I. Energy measurement of relativistic electron beams by laser Compton scattering. Physical Review E. 1996;**54**(5):5657-5663. DOI: 10.1103/PhysRevE.545657

[54] Baylac M, Burtin E, Cavata C, Escoffier S, Frois B, Lhuillier D, et al. First electron beam polarization measurements with a Compton polarimeter at Jefferson Laboratory. Physics Letters B. 2002;**539**:8-12. DOI: 10.1016/S0370-2693(02)02091-9

[55] Omori T, Fukuda M, Hirose T, Kurihara Y, Kuroda R, Nomura M, et al. Efficient propagation of polarization from laser photons to positrons through Compton scattering and electron-positron pair creation. Physical Review Letters. 2006;**96**(11):114801. DOI: 10.1103/PhysRevLett.96.114801

[56] Kikuzawa N, Hajima R, Nishimori N, Minehara E, Hayakawa T, Shizuma T, et al. Nondestructive detection of heavily shielded materials by using nuclear resonance fluorescence with a laser-Compton scattering γ-ray source. Applied Physics Express. 2009;**2**: 036502. DOI: 10.1143/APEX.2.036502

[57] Hajima R, Kikuzawa N, Nishimori N, Hayakawa T, Shizuma T, Kawase K, et al. Detection of radioactive isotopes by using laser Compton scattered gamma-ray beams. Nuclear Instruments and Methods in Physics Research Section A: Accelerators, Spectrometers, Detectors and Associated Equipment. 2009;**608**:S57-S61. DOI: 10.1016/j.nima.2009.05.063

[58] Hayakawa T, Ohgaki H, Shizuma T, Hajima R, Kikuzawa N, Minehara E, et al. Nondestructive detection of hidden chemical compounds with laser Compton-scattering gamma rays. Review of Scientific Instruments. 2009;**80**(4):045110. DOI: 10.1063/1.3125022

[59] Albert F, Anderson SG, Anderson GA, Betts SM, Gibson DJ, Hagmann CA, et al. Isotope-specific detection of low-density materials with laser-based monoenergetic gamma-rays. Optics Letters. 2010;**35**(3):354-356. DOI: 10.1364/OL.35.000354

[60] Albert F, Anderson SG, Gibson DJ, Hagmann CA, Johnson MS, Messerly M, et al. Characterization and applications of a tunable, laser-based MeV-class Compton-scattering γ-ray source. Physical Review Accelerators and Beams. 2010;**13**(7):070704. DOI: 10.1103/PhysRevSTAB.13.070704

[61] Ohgaki H, Daito I, Zen H, Kii T, Masuda K, Misawa T, et al. Nondestructive inspection system for special nuclear material using inertial electroelastic confinement fusion neutrons and laser Compton scattering gamma-rays. IEEE Transactions on Nuclear Science. 2017;**64**(7):1635-1640. DOI: 10.1109/TNS.2017.2652619

[62] Jochmann A, Irman A, Bussmann M, Couperus JP, Cowan TE, Debus AD, et al. High resolution energy-angle correlation measurement of hard X rays from laser-Thomson backscattering. Physical Review Letters. 2013;**111**(11):114803. DOI: 10.1103/PhysRevLett.111.114803

[63] Ta Phuoc K, Rousse A, Pittman M, Rousseau JP, Malka V, Fritzler S, et al. X-ray radiation from nonlinear Thomson scattering of an intense femtosecond laser on relativistic electrons in helium plasma. Physical Review Letters. 2003;**91**(19):195001. DOI: 10.1103/PhysRevLett.91.195001

[64] Schwoerer H, Liesfeld B, Schlenvoigt H-P, Amthor K-U, Sauerbrey R, Thomson-backscattered X. Rays from laser-accelerated electrons. Physical Review Letters. 2006;**96**(1):014802. DOI: 10.1103/PhysRevLett.96.014802

[65] Ta Phuoc K, Corde S, Thaury C, Malka V, Tafzi A, Goddet JP, et al. All-optical Compton gamma-ray source. Nature Photonics. 2012;**6**(5):308-311. DOI: 10.1038/NPHOTON.2012.82

[66] Chen S, Powers ND, Ghebregziabher I, Maharjan CM, Liu C, Golovin G, et al. MeV-energy X rays from inverse Compton scattering with laser-wakefield accelerated electrons. Physical Review Letters. 2013;**110**(15):155003. DOI: 10.1103/PhysRevLett.110.155003

[67] Powers ND, Ghebregziabher I, Golovin G, Liu C, Chen S, Banerjee S, et al. Quasi-monoenergetic and tunable X-rays from a laser-driven Compton light source. Nature Photonics. 2014;**8**:28-31. DOI: 10.1038/nphoton.2013.314

[68] Sarri G, Corvan DJ, Schumaker W, Cole JM, Di Piazza A, Ahmed H, et al. Ultrahigh brilliance multi-MeV γ-ray beams from nonlinear relativistic Thomson scattering. Physical Review Letters. 2014;**113**(22):224801. DOI: 10.1103/PhysRevLett.113.224801

[69] Miura E, Ishii S, Tanaka K, Kuroda R, Toyokawa H. X-ray pulse generation by laser Compton scattering using a high-charge, laser-accelerated, quasi-monoenergetic electron beam. Applied Physics Express. 2014;**7**:046701. DOI: 10.7567/APEX.7.046701

[70] Tsai H-E, Wang X, Shaw JM, Li Z, Arefiev AV, Zhang X, et al. Compact tunable Compton x-ray source from laser-plasma accelerator and plasma mirror. Physics of Plasmas. 2015;**22**(2):023106. DOI: 10.1063/1.4907655

[71] Khrennikov K, Wenz J, Buck A, Xu J, Heigoldt M, Veisz L, Karsch S. Tunable all-optical quasimonochromatic Thomson X-ray source in the nonlinear regime. Physical Review Letters. 2015;**114**(19):195003. DOI: 10.1103/PhysRevLett.114.195003

[72] Liu C, Golovin G, Chen S, Zhang J, Zhao B, Haden D, et al. Generation of 9 MeV γ-rays by all-laser-driven Compton scattering with second-harmonic laser light. Optics Letters. 2014;**39**(14):4132-4135. DOI: 10.1364/OL.39.004132

[73] Shaw JM, Bernstein AC, Hannasch A, LaBerge M, Chang Y-Y, Weichman K, et al. Generation os tens-of-MeV photons by Compton backscatter from laser-plasma-accelerated GeV electrons. AIP Conference Proceedings. 2017;**1812**:100012. DOI: 10.1063/1.4975910

[74] Scoby CM. Adapting High Brightness Relativistic Electron Beams for Ultrafast Science [Dissertation]. Los Angeles, California, USA: The University of California; 2013. 165 p. Available from: http://escholarship.org/uc/item/11p8s2h1

[75] Lundh O, Lim J, Rechatin C, Ammoura L, Ben-Ismaïl A, Davoine X, et al. Few-femtosecond, few kiloampere electron bunch produced by a laser-plasma accelerator. Nature Physics. 2011;**7**:219-222. DOI: 10.1038/nphys1872

[76] Cianchi A, Anania MP, Bisesto F, Castellano M, Chiadroni E, Pompili R. Observations and diagnostics in high brightness beams. Nuclear Instruments and Methods in Physics Research Section A: Accelerators, Spectrometers, Detectors and Associated Equipment. 2016;**829**:343-347. DOI: 10.1016/j.nima.2016.03.076

[77] Gizzi LA, Benedetti C, Cecchetti CA, Di Pirro G, Gamucci A, Gatti G, et al. Laser-plasma acceleration with *FLAME* and *ILIL* ultraintense lasers. Applied Sciences. 2013;**3**:559-580. DOI: 10.3390/app3030559

[78] Zou JP, Le Blanc C, Papadopoulos DN, Chériaux G, Georges P, Mennerat G, et al. Design and current progress of the Apollon 10 PW project. High Power Laser Science and Engineering. 2015;**3**:e2. DOI: 10.1017/hpl.2014.41

[79] Agrawal GP. Nonlinear Fiber Optics. 4th ed. San Diego: Academic Press; 2006. 529 p. DOI: 10.1016/B978-012369516-1/50000-5

[80] Benedetti C, Schroeder CB, Esarey E, Leemans WP. Plasma wakefields driven by an incoherent combination of laser pulses: A path towards high-average power laser-plasma accelerators. Physics of Plasmas. 2014;**21**(5):056706. DOI: 10.1063/1.4878620

[81] Vargas M, Schumaker W, He Z-H, Behm K, Chvykov V, Hou B, et al. Improvements to laser wakefield accelerated electron beam stability, divergence, and energy spread using three-dimensional printed two-stage gas cell targets. Applied Physics Letters. 2014; **104**(17):174103. DOI: 10.1063/1.4874981

[82] He Z-H, Hou B, Nees JA, Easter JH, Faure J, Krushelnick K, Thomas AGR. High repetition-rate wakefield electron source generated by few-millijoule, 30 fs laser pulses on a density downramp. New Journal of Physics. 2013;**15**(5):053016. DOI: 10.1088/1367-2630/15/5/053016

[83] He Z-H, Hou B, Lebailly V, Nees1 JA, Krushelnick K, Thomas AGR. Coherent control of plasma dynamics. Nature Communications. 2015;**6**:7156. DOI: 10.1038/ncomms8156

[84] Zhu W, Palastro JP, Antonsen Jr TM. Studies of spectral modification and limitations of the modified paraxial equation in laser wakefield simulations. Physics of Plasmas. 2012;**19**(3):033105. DOI: 10.1063/1.3691837

[85] Lifschitz AF, Davoine X, Lefebvre E, Faure J, Rechatin C, Malka V. Particle-in-cell modelling of laser–plasma interaction using Fourier decomposition. Journal of Computational Physics. 2009;**228**:1803-1814. DOI: 10.1016/j.jcp.2008.11.017

[86] Lehe R, Lifschitz A, Thaury C, Malka V, Davoine X. Numerical growth of emittance in simulations of laser-wakefield acceleration. Physical Review Accelerators and Beams. 2013;**16**(2):021301. DOI: 10.1103/PhysRevSTAB.16.021301

[87] Jackson JD. Classical Electrodynamics. 3rd ed. New York: Wiley; 1999. 832 p

[88] Balakin AA, Litvak AG, Mironov VA, Skobelev SA. Compression of femtosecond petawatt laser pulses in a plasma under the conditions of wake-wave excitation. Physical Review A. 2013;**88**(2):023836. DOI: 10.1103/PhysRevA.88.023836

[89] Schreiber J, Bellei C, Mangles SPD, Kamperidis C, Kneip S, Nagel SR, et al. Complete temporal characterization of asymmetric pulse compression in a laser wakefield. Physical Review Letters. 2010;**105**(23):235003. DOI: 10.1103/PhysRevLett.105.235003

[90] Weingartner R, Fuchs M, Popp A, Raith S, Becker S, Chou S, et al. Imaging laser-wakefield-accelerated electrons using miniature magnetic quadrupole lenses. Physical Review Accelerators and Beams. 2011;**14**(5):052801. DOI: 10.1103/PhysRevSTAB.14.052801

[91] Grigsby FB, Dong P, Downer MC. Chirped-pulse Raman amplification for two-color, high-intensity laser experiments. Journal of the Optical Society of America B. 2008;**25**(3): 346-350. DOI: 10.1364/JOSAB.25.000346

[92] Sanders JC, Zgadzaj R, Downer MC. Two-color terawatt laser system for high-intensity laser-plasma experiments. AIP Conference Proceedings. 2012;**1507**:882-886. DOI: 10.1063/1.4773816

[93] Vicario C, Shalaby M, Konyashchenko A, Losev L, Hauri CP. High-power femtosecond Raman frequency shifter. Optics Letters. 2016;**41**(20):4719-4722. DOI: 10.1364/OL.41.004179

Section 3

Radiation Safety Aspects

Chapter 5

Nuclear Safety Study of High Energy Heavy-ion Medical Accelerator Facility

Oyeon Kum

Additional information is available at the end of the chapter

http://dx.doi.org/10.5772/intechopen.70827

Abstract

During beam operation in heavy-ion medical accelerator facilities, radiological problems may arise during normal operation and by accidental loss in the beam system. This study emphasizes the nuclear safety aspects in designing a heavy-ion medical accelerator facility, with preliminary design concepts to accommodate a new synchrotron medical accelerator with a maximum energy of 430 MeV/u carbon ions. The beam loss points and irradiation rooms, which are potential hazardous areas of radiation exposure, are described for radiation shielding and activation simulations. Shielding simulations were performed according to the NCRP 147 recommendations, including skyshine and groundshine in a conservative manner with the occupancy factor of 1.0 and workload of 100%. The carbon 12 ions of energy 430 MeV/u generate radioactive isotopes as they interact with surrounding air and accelerator system components during transmission. The activation phenomena in air, cooling water, underground soil and ground water, and typical accelerator component materials such as iron and copper were estimated in detail. Nuclear safety simulations were performed by using the combination of MCNPX2.7.0 and the CINDER'90 codes. Thus, this report will provide a useful guide for estimating radiological impacts and allow optimal design of heavy ion medical accelerator facilities with high safety standards.

Keywords: heavy-ion medical accelerator facility, optimized design, nuclear safety, radiation shielding, activation protection

1. Introduction

Safety assessment is a systematic and comprehensive methodology to evaluate risks associated with a complex technological entity such as a high-energy heavy-ion medical accelerator facility. In general, National Nuclear Energy Safety Board formulates safety requirements for nuclear and radiation facilities to analyze their safety over five stages of their life-cycle: siting, design, construction, commissioning, operation and decommissioning. Because safety analysis aims to achieve completeness in defining possible mishaps, deficiencies, and plant vulnerabilities, producing a balanced picture of significant safety issues across a broad spectrum is important. Thus, two or three levels of safety assessment over the years are accepted as an international standard [1].

In this study, radiation protection and safety design concepts are introduced by taking an example of a new heavy-ion medical accelerator facility in Korea [2]. This report emphasizes only nuclear safety aspects such as radiation shielding and activation analysis for the preliminary design of the facility. For simulations of radiation shielding and activation analysis, detailed information on the beam parameters such as beam intensities along the different accelerator beam lines, transport coefficients, beam loss points, and operating times is required. The Monte Carlo code, MCNPX2.7.0 [3] and transmutation code, CINDER'90 [4] were used as simulation tools.

The MCNPX2.7.0 code is widely used and is the latest version in the MCNPX code development series. The MCNPX2.7.0 code includes nuclear data tables to transport protons, physics models to transport 30 or more additional particle types such as deuterons, tritons, alphas, pions, muons, and additional physics models to transport neutrons and protons when no tabular data are available. The mix and match model used in MCNPX2.7.0 makes the interface between table physics and model physics seamless. On the other hand, the transmutation code, CINDER'90, is not widely used although its development history is over five decades. It solves the Bateman equation directly by using Laplace transformation for 3400 nuclides including ground state and isomeric states. There are 98 nuclides with fission paths; 58 via spontaneous fission (SF) and 67 via neutron-induced fission (n, f). However, the combination of MCNPX and CINDER'90 provides accurate results for many different radiation protection-related problems. A brief description on a new synchrotron accelerator in design phase is provided in Section 2. In Section 3, radiation shielding methodologies for the facility design is introduced. In Section 4, activation estimations for air, cooling water, ground water and underground soil, and accelerator materials are provided. Other radiological effects in surrounding environments are also discussed. Conclusions are given in Section 5.

2. A new synchrotron medical accelerator in design phase

The entire accelerator system consists of the injector, circular accelerator (synchrotron), and high-energy beam transport (HEBT). For medical accelerator systems, the end of the beam transport is connected to the medical treatment system (or beam application system) which is not part of the accelerator system. **Figure 1** shows schematic accelerator building and a new medical synchrotron accelerator system, currently in design phase, with two external injectors and the high-energy beam transport to the four irradiation rooms.

The injector is the functional and spatial representation of a subgroup of the accelerator system where the ions are produced from suitable sources and pre-accelerated under DC fields, and are entered into the synchrotron with high-frequency fields of the resonant structures. Two

http://dx.doi.org/10.5772/intechopen.70827

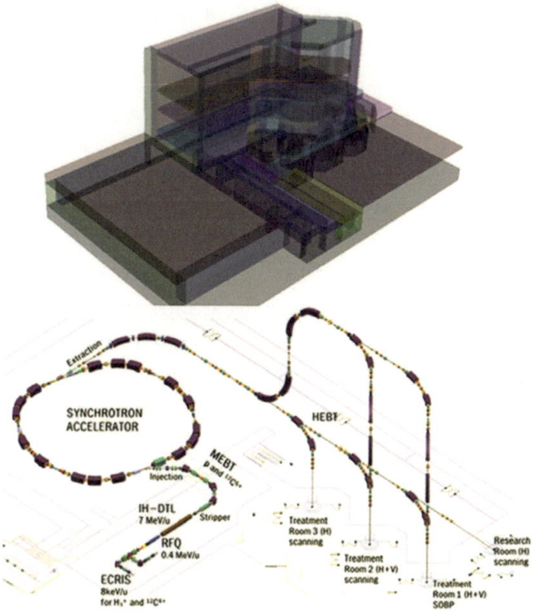

Figure 1. Schematic layout of the building (top) and the subdivision of the accelerator with three main systems (bottom): injection, synchrotron, and high-energy beam transport.

injector inlets can generate two different ions such as carbon ion and proton (or oxygen), and pass them alternatively to the synchrotron. The switching between different ions must be fast for practical purposes. Two ion beams are accelerated to the typical state of charge of energy of 10 keV per nucleon under the DC voltage of a maximum of 30 kV. Ion sources are produced in the type of "electron cyclotron resonance". Two sources are used for the medical use: carbon ions in the form of 12C4+ and hydrogen ions in the form of H3+.

The beam line of medium-energy beam transport (MEBT) starts at the exit of the low-energy beam transport and extends to the electrostatic injection septum in the synchrotron (see **Figure 1**). In this system, the particle energy is accelerated up to 7 MeV/u. Remaining electrons of the previously generated and pre-accelerated H3+ or 12C4+ ions are stripped in the film of the stripper. H3+ ion is divided into three individual protons in the stripper film. 12C6+ ions are also produced at the exit of the film. This alters the particle's charge to mass ratio: from formerly q/m = 1/3 for both ions to q/m = 1/1 for protons and q/m = 1/2 for carbon ions. The electrostatic injection septum is the last element of the medium-energy beam transport and is already part of the synchrotron. This element is divided into two parts: a field-free part where the circulating beam passes and the second part with an electrostatic field passing through the iron and, thus, drawn to the nominal orbit in the synchrotron. The scattered particles with a kinetic energy of 7 MeV/u are, however, negligible in terms of radiation protection.

At the end of the medium-energy beam transport, the synchrotron operates. It accumulates the injection process, which takes several rounds. The acceleration of the particles from the injection energy to the final energy is achieved in the synchrotron. After the beam obtains a desirable energy, beam extraction is guided from the synchrotron to the HEBT. The layout of the synchrotron with the circumference of 77.648 m and identifies some important elements such as the electrostatic injection septum, the electrostatic extraction septum, the magnetic extraction septum, the sextupole magnets to control the chromaticity, the RF cavity, and sextupole magnets for the resonant extraction.

In order to keep the beam in the closed ring of synchrotron, nearly circular orbit, 16 identical dipole magnets are used with a length of about 1.7 m, and deflect the beam by 22.5°, respectively. It is at the maximum extraction energy of carbon ions of 430 MeV/u and at the required magnetic field of 1.5 T. In this critical condition, the individual particles of the beam does not diverge from its orbit with the help of 24 main quadrupole magnets and 2 other special quadrupole magnets (1 with air coil and a 45° rotated around the longitudinal axis). The maximum required magnetic field gradient is approximately 4 T/m. In addition, four sextupole magnets are used to compensate for the chromaticity and a special sextupole magnet is installed in a dispersion-free section of the synchrotron for the resonance extraction.

Fast solenoids activate the beam stop and the beam absorber is required to absorb the beam within 50 μs, if no extraction is required. Correction dipole magnets are used individually to control the correction of small errors in the orbit. In the injection channel, around 75% of the particles coming from the injector are lost. A particle loss of about 5% at the electrostatic septum and at the three magnetic septa is also expected. When circulating radiation beam in the synchrotron need not be extracted completely, two beam stopper magnets are used to block the beam completely. Therefore, major radiation sources of maximum energy, 430 MeV/u, are generated in the synchrotron.

The beam transport lines from the synchrotron to the beam application system in each irradiation rooms are called the HEBT. High-energy beam transport begins with the extraction line from the synchrotron and then divides into the individual beam delivery lines to the irradiation rooms. The breaker is housed in the first portion of the extraction line to steer the beam extracted from the synchrotron to the direction of the beam absorber, which is located at the end of the extraction system. The beam supply lines are designed for each individual exposure area, and form a beam size in the horizontal and vertical plane at a ratio of 1:1.55 and 1:1, respectively. Each line has a beam stopper which can block the beam line completely for safety purposes.

3. Shielding simulations

3.1. Beam source points

Based on user requirements, the gross intensity of the particle beams along the accelerator chain is created at the beginning of the project. This serves as a basis for the design of the accelerator facility. In this section, the technical aspects of assessing radiation protection are described. In the assessment, it is assumed that the maximum radiation intensities to prevent any restriction for future operation. An improved knowledge of beam parameters enables both reduction of annual beam intensities and accurate determination of relative beam loss during operation. **Figure 2** shows a graphical overview of beam loss points and four irradiation rooms. Typical energies at each loss point are also shown.

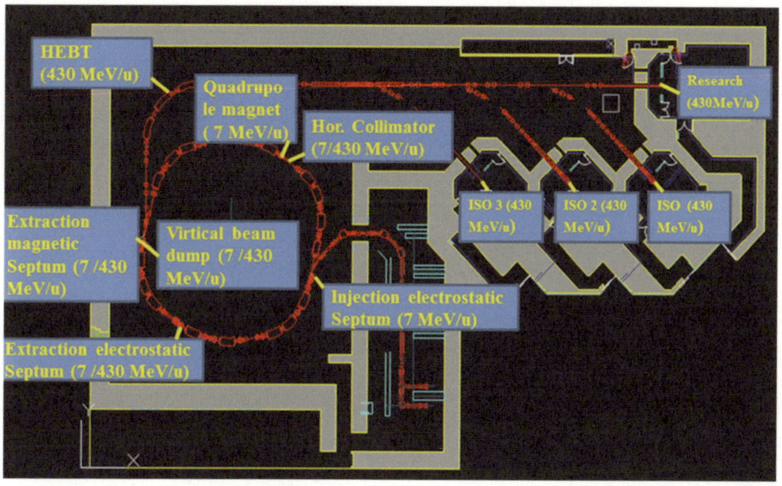

Figure 2. Graphical overview of beam loss points and four irradiation rooms. Typical energies at each loss point are shown.

In heavy-ion medical accelerator facilities, the beam loss points are potential hazardous areas of radiation exposure because the primary and the secondary particles generated by the collisions between the lost beam and the accelerator structures produce abundance gamma rays and other radioactive nuclides. In shielding and radiative simulations, the lost beam intensities estimated by the beam physics experts who are in charge of the conceptual design of the accelerator are important parameters. The beam loss point is divided into several categories according to the energy. For a conservative assessment, the maximum beam energy (430 MeV/u) is always used for simulations. In addition, lost beam intensities are also assumed to be maximal: 1×10^9 particles per second (pps) for beam loss points and irradiation rooms. Operation time per year is assumed to be 2000 h/year (8 h/day $\times$ 250 days/year). Thus, the simulation results with maximum energy and intensity can give higher degrees of freedom for future operation policy.

3.2. Dose calculation sites

Shielding simulations were performed according to the NCRP 147 [5] recommendations. However, for a conservative estimation, an occupancy factor and workload of 1.0 and 100% were assumed, respectively. Dose evaluation points are located 30 cm away from the shield wall and 50 cm above from the floor surface for the same level and upper level of the source point, respectively, as shown in **Figure 3**.

Accuracy verification of the MCNPX2.7.0 code was performed by comparing the results of the FLUKA2011 [6] code with the benchmark problems. The results were reported elsewhere [7]. Relative iso-dose color levels for the entire accelerator facility are shown in **Figure 4**. For the accurate evaluation, a total of 22 sites were selected for the dose calculation as shown in **Figure 5**. The point numbers in **Figure 4** are named as follows: (1) synchrotron vault entrance; (2) front side power station room; (3 and 4) left outside wall of the synchrotron vault

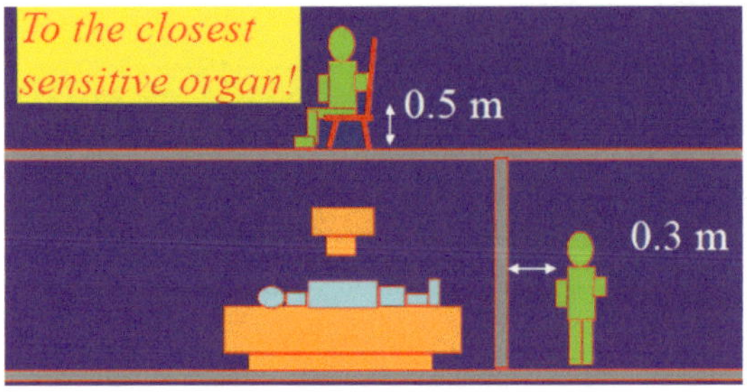

Figure 3. Dose evaluation points recommended by NCRP 147.

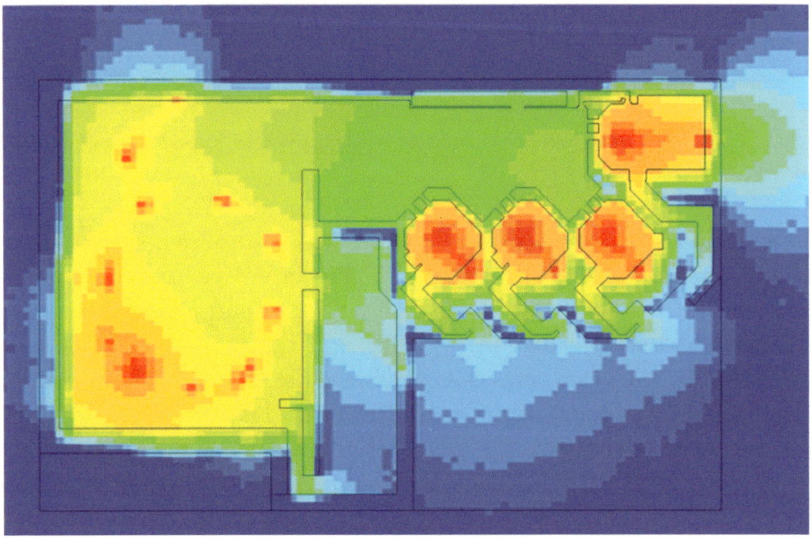

Figure 4. Relative iso-dose color levels in the whole accelerator facility.

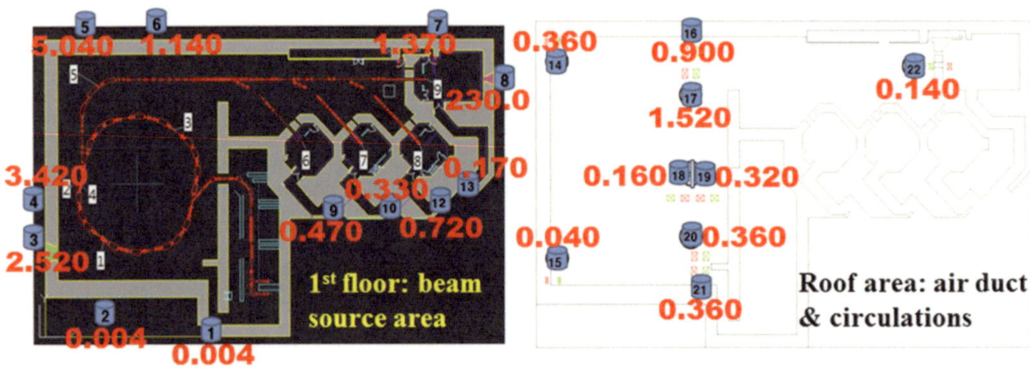

Figure 5. Dose estimation points and corresponding total effective dose per year (unit: mSv/year) red color (or dark grey). Computing errors are less than 4%.

(front and back side, respectively); (5–7) outside of back side wall (left, in, and right side, respectively); (8) outside of right wall (back side); (9, 10, and 12) entrances of treatment rooms (no. 3, 2, and 1, respectively); (13) entrance of research room; (14 and 15) outlet of air duct (no. 1 and 2, respectively); (16) inlet of air circulation 3-1; (17) outlet of air circulation 3-2; (18) inlet of air circulation 4-1; (19) outlet of air circulation 4-2; (20) inlet of air circulation 5-1; (21) outlet of air circulation 5-2; (22) outlet of air duct 6. The estimated effective doses are written in red color (or dark grey) as shown in **Figure 5**. Point 11 is missing in **Figure 5**.

3.3. Skyshine

Skyshine is described as the ionizing radiation emitted by an accelerator facility, reaching the facility's surroundings not directly, but indirectly through reflection and scattering at the atmosphere back to the earth's surface. When the shielding barrier around the source of radiation is not enough at the top, skyshine can happen. In general, skyshine is also described as the radiation reflected off the ceiling inside an accelerator facility. The intensity of radiation measured at the surface surrounding the facility increases with growing distance from the shielding barrier. The skyshine model was accounted directly in the simulation by including a large volume of outside air with a height of 500 m and a width/length of 100 m as shown in **Figure 6**, including measuring points and the calculated results. Outside 10% of sidewall thicknesses are defined as complete beam absorption area so that the beam never reaches outside through the sidewalls, but inside back scattering effects are included. The total effective doses (neutron + gamma) at the height of 120 cm are written on the measuring points (red color) in the left side of **Figure 6**.

3.4. Groundshine

Groundshine describes the radiation leakage through the concrete floor, which is scattered in the underground soil and then rises to the ground, affecting the surrounding environment. This effect can occur when the thickness of the floor concrete is not thick enough. Generally, the longer the distance of radiation traveled, the higher the probability of interacting with the underground soil. Thus, this radiation gradually loses more energy. Therefore, a conservative

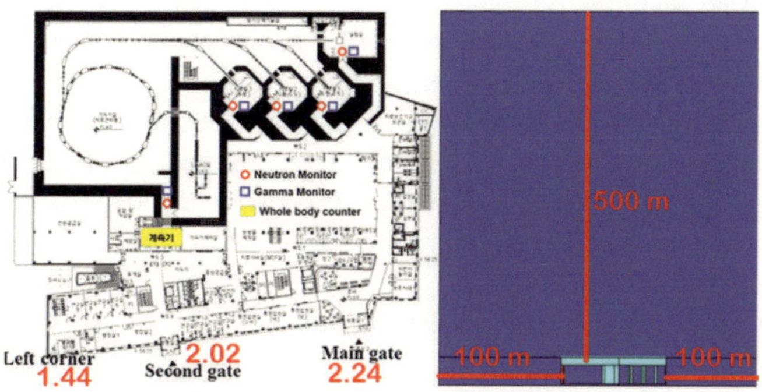

Figure 6. Approximate three measuring points (left) and a simplified skyshine simulation model (right). Additional air volume of $100 \times 100 \times 500$ m^3 is included in the model. Total (neutron + photon) dose rates (unit: μSv/year) at the height of 120 cm are written on the measuring points in red color (or dark grey). Maximum relative error is about 4%.

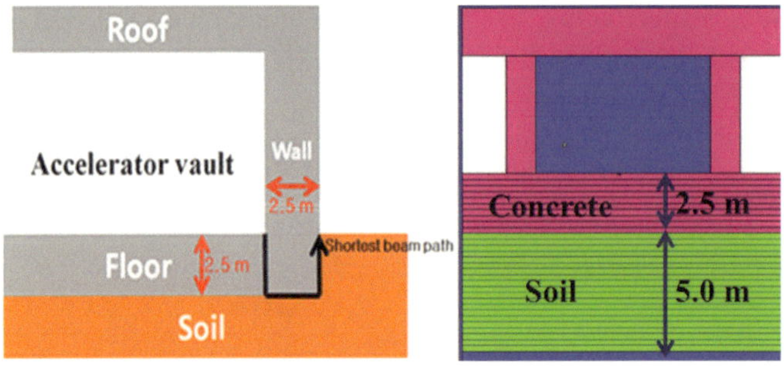

Figure 7. Groundshine calculation model. The left side shows the shortest path of radiation leakage to outside. The right side is the corresponding simulation model with 2.5 m of the concrete and an additional 5.0 m of soil.

model can be found with the shortest travel distance, as shown in **Figure 7**, because the radiation traveled to the shortest distance will have the greatest effect on the groundshine. As shown in the left of **Figure 7**, the shortest distance follows the black arrow, 2.5 m of the concrete floor and an additional 5.0 m of soil. Corresponding simplified model is shown in the right of **Figure 7**. The total dose rate of groundshine was calculated to be 1.39 μSv/year.

4. Activation assessments

High-energy carbon ion beams generate radioactive isotopes as they interact with the surrounding air and accelerator system components during transmission. Since the radioactive isotopes increase the radiation risk, the activation phenomenon is one of the most important issues to be evaluated in the high-energy particle accelerator facility. In this study, the activation problem of the medical accelerator facility was evaluated by using the combination of MCNPX2.7.0 and the CINDER'90 codes.

4.1. Air activation

This section calculates the amount of radioactive isotopes generated by different beam usage scenarios, and analyses its impacts on the environment and also on the staffs and visitors. The scenario used in this calculation assumes the most dangerous and conservative cases that can occur in the treatment rooms and the accelerator hall. Assessment is performed by using easy-to-understand generic models. Safety analysis is based on the standard issued by the National Nuclear Safety Commission and the Radiation Safety Commission. Although the maximum beam energy is not always used in the patient treatment and experiments, the maximum energy scenario is used to obtain conservative and safe results.

4.1.1. Air activation model of treatment room or research room

To model the treatment rooms, a simple geometry with an internal space of 5 m × 8 m × 10 m surrounded by concrete walls of 1 m thickness is created. The concrete wall thickness of 1 m is enough to guarantee back scattering and secondary particle generation. A typical patient is modeled by the 30 cm (diameter) × 35 cm (length) cylinder phantom with human tissue-like substance (hereinafter referred to as "tissue phantom") which has tissue equivalent material defined in ICRU-44 [8]. **Figure 8** shows the arrangement of treatment room and phantom. The volume of this model is similar to that of real treatment room.

The use of a tissue phantom, not a pure water phantom, is intended to perform a more realistic radioactive simulation. The carbon ion beam is incident on the central axis of the cylindrical phantom from 1 m behind. This separation generally describes the distance between the patient's body and the tip of the beam nozzle when the treatment is performed. Since the maximum penetration depth of the human bone with 430 MeV/u of carbon energy is about 27 cm, the current set can be regarded as a conservative model in which the neutrons are released to the outside and the surrounding air is maximally activated. The generation of radionuclides is attributed to the two types of beams: (1) primary carbon beam and (2) secondary beam caused by collision of a carbon beam with the tissue phantom.

The difference between research and treatment rooms is the target material. The target substances used in the laboratory may vary greatly. They could be the biological organic materials, various inorganic minerals, or composite objects. However, in this study, steel was used as the most popular material. The shape of the target is cylindrical with a relatively large diameter of 10 cm and a length of 10 cm, which is shorter than the attenuation length of 430 MeV/u carbon ions in steel and, thus, expected to produce both a low-energy primary carbon ions and relatively large number of secondary particles. In this case, spallation generated by the low-energy primary carbon beam passing through the target can also be expected.

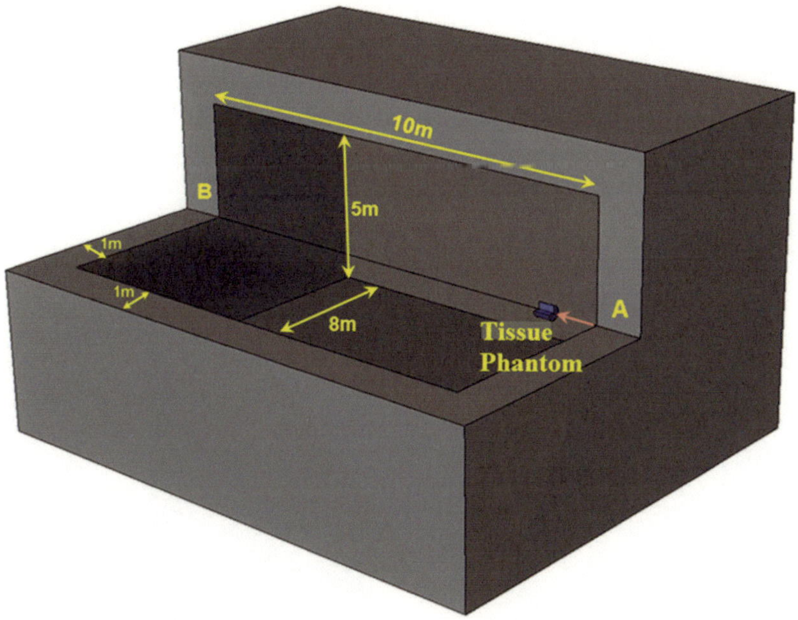

Figure 8. Generic model of treatment or research room.

4.1.2. Air activation model of accelerator hall

The entire volume of accelerator hall cannot be simulated directly because it is too big to simulate exactly. In this study, small volume model is used for simulation, and the results will be extrapolated to the entire volume. The model volume of the accelerator hall is 5 m × 5 m × 10 m with one-meter thick concrete walls. The shape of the target is cylindrical with a relatively large diameter of 10 cm and a length of 20 cm. To compare the difference between spallation generated by the primary carbon beam and the absorption by thermal neutrons, two target positions are selected. The first target position (T1) is 1 m away from the primary beam. This target may generate a comparatively large number secondary beams and thermal scattered beams. On the other hand, second target position (T2) is 9 m away from the carbon beam source. In this simulation, target T1 is removed so that this target may generate more primary collision effects (compared to those of T1 target) because the beam passes through the entire room before it hits target T2. Note that these target positions are arranged centrally in the room and 1 m away from the walls A and B, respectively, so that the air of the room (1 m thick) surrounded by concrete walls accounts for the back scattering of secondary particles. The generic geometry model is shown in **Figure 9**.

The irradiation scenario for each irradiation room has the same maximum energy of 430 MeV/u, maximum beam intensity of 1×10^9 pps, no ventilation, but only different beam irradiation times. Beam irradiation modes are divided into two cases: normal operation and accidental case. Irradiation times for normal operation are 120 and 300 s for treatment room and research room, respectively. Irradiation times for accidents are 1, 1, and 24 h for treatment room, research room, and accelerator hall, respectively. Accelerator hall assumes 1 h stay just after 24 h irradiation.

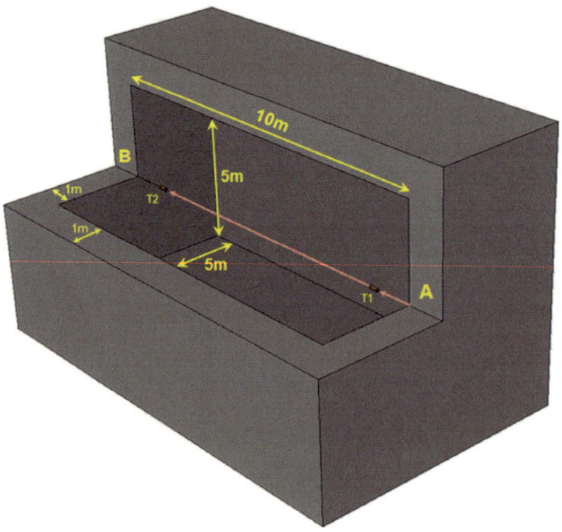

Figure 9. Generic model of accelerator hall.

4.1.3. Results and discussion

4.1.3.1. Spallation and neutron absorption effects

Depending on the target position, spallation and neutron absorption effects show conspicuous differences as shown in **Figure 10** and **Table 1**. H-3, Be-7, and C-11 nuclides, which are typical spallation products, are generated more at target 2 which was hit after 9 m travel of primary carbon ion beam in the air. On the other hand, Cl-38, Ar-37, and Ar-39 nuclides, typical thermal neutron absorption products, are generated more at target T1, which allows more secondary particles to interact with the air.

4.1.3.2. Radiological impact for surrounding environments

Radiological effects for surrounding environments due to the beam loss in the accelerator hall were analyzed by a 1 year ratio of specific activity relative to the regulation limit. For conservative analysis, regulation limits of S-38 and Cl-34 m are both 0.2, which is the smallest value for similar mass isotope, Fe-60. No regulation limits are given for the S-38 and Cl-34 m. This may make the total sum slightly bigger. The exact regulation limits for these two isotopes are calculated in Appendix A. Analytically calculated values are shown in parenthesis, which reduce the total sum greatly (see **Table 2**).

4.1.3.3. Air volume dependence of radiological impact

Air volume dependence of radiological impact is analyzed using two different air volume models: 5 m × 5 m × 10 m (=250 m^3) and 10 m × 10 m × 10 m (=1000 m^3). Total sums of the ratios are 0.537 and 0.763 for small and large volumes, respectively. A volume increase by a factor of four makes the total sum ratio to increase about 42%. Thus, the exponent, ln(1.42)/ln(4), is approximately 0.253 and volume dependence equation is as follows:

$$\text{Radiological impact in arbitrary volume} = \text{Radiological impact in control volume} \times (Va/Vc)^{0.253} \quad (1)$$

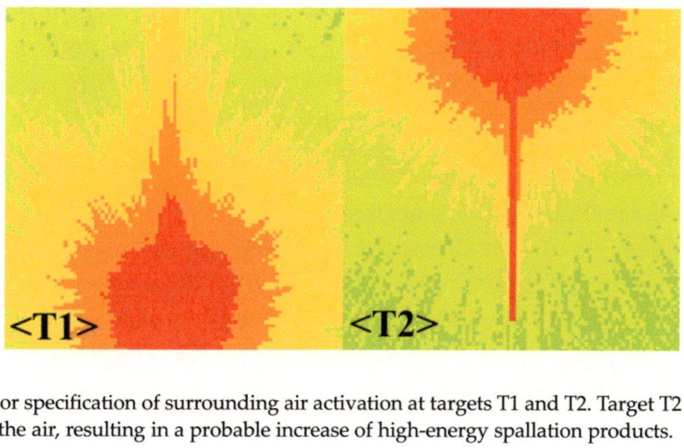

Figure 10. Relative color specification of surrounding air activation at targets T1 and T2. Target T2 shows long passage of primary carbon ion in the air, resulting in a probable increase of high-energy spallation products.

Nuclide	Half-life (s)	1 h irradiation		
		T1	T2	T1/T2
H 3	3.89E + 08	2.24E − 08	9.24E − 08	2.43E − 01
Be 7	4.61E + 06	3.32E − 07	5.86E − 06	5.65E − 02
C 11	1.22E + 03	6.04E − 04	6.19E − 03	9.76E − 02
N 13	5.98E + 02	1.37E − 03	3.86E − 03	3.54E − 01
O 14	7.06E + 01	5.00E − 05	3.52E − 04	1.42E − 01
O 15	1.22E + 02	6.16E − 04	1.57E − 03	3.92E − 01
Na 24	5.39E + 04	2.76E − 47	2.72E − 07	1.01E − 40
Al 28	1.34E + 02	4.00E − 06	1.00E − 05	4.00E − 01
Al 29	3.94E + 02	5.17E − 24	3.99E − 06	1.29E − 18
Si 31	9.44E + 03	1.39E − 06	9.29E − 07	1.50E + 00
P 32	1.23E + 06	3.24E − 08	4.45E − 08	7.27E − 01
P 33	2.19E + 06	9.11E − 09	1.37E − 08	6.67E − 01
P 35	4.73E + 01	2.00E − 06	1.99E − 06	1.01E + 00
S 35	7.56E + 06	5.93E − 09	5.27E − 09	1.13E + 00
S 37	3.03E + 02	6.00E − 06	8.00E − 06	7.50E − 01
S 38	1.02E + 04	8.67E − 07	8.67E − 07	1.00E + 00
Cl 38	2.23E + 03	4.07E − 05	2.86E − 05	1.42E + 00
Cl 39	3.34E + 03	2.74E − 05	2.32E − 05	1.18E + 00
Ar 37	3.03E + 06	3.79E − 08	3.13E − 08	1.21E + 00
Ar 39	8.49E + 09	4.06E − 11	3.38E − 11	1.20E + 00

Table 1. Specific activities for the targets T1 and T2. Ratio of the products is shown in the last column.

Here Vc is control volume and Va is arbitrary volume.

Specific activity and their ratios are shown in **Table 3**. This volume correction equation makes the accelerator hall calculation easy because the real analysis can be done on a small model.

4.1.3.4. Radiological impacts of medical staffs by inhalation

Inhalation dose estimation (committed effective dose) relies on the characterization of the airborne radioactive material and human body response to inhaled radionuclides. Inhaled radionuclides are either exhaled or deposited in various regions and segments of the human respiratory tract (e.g., see ICRP1994 [9] and 1979 [10]). Because radioactive decay within internal organs can continue for several years or decades after the original intake of radioactive nuclides, internal dose calculations are performed for a commitment period that is typically applied as the receipt of a single dose in the year (unit: Sv/year) of intake that equals the accumulated dose over the commitment period, which is a 50-year commitment period for

Nuclide	Half-life (s)	Specific activity (Bq/m^3)	Regulation limit (Bq/m^3)	Ratio
H 3	3.89E + 08	2.78E − 03	2.00E + 03	1.39E − 06
Be 7	4.61E + 06	7.70E − 02	1.00E + 03	7.70E − 05
C 11	1.22E + 03	9.73E + 00	2.00E + 04	4.87E − 04
N 13	5.98E + 02	5.25E + 00	6.00E + 03	8.74E − 04
Si 31	9.44E + 03	8.15E − 02	9.00E + 02	9.05E − 05
Si 32	4.17E + 09	9.12E − 08	6.00E − 01	1.52E − 07
P 32	1.23E + 06	3.38E − 03	2.00E + 01	1.69E − 04
P 33	2.19E + 06	1.56E − 03	5.00E + 01	3.12E − 05
S 35	7.56E + 06	1.51E − 03	5.00E + 01	3.01E − 05
S 38	1.02E + 04	2.58E − 02	2.00E − 01 (423)	1.29E − 01 (6.10E − 5)
Cl 34 m	1.93E + 03	3.04E − 02	2.00E − 01 (17)	1.52E − 01 (1.79E − 3)
Cl 38	2.23E + 03	1.26E + 00	1.00E + 03	1.26E − 03
Cl 39	3.34E + 03	1.44E + 00	1.00E + 03	1.44E − 03

Table 2. One year ratio of specific activity divided by the regulation limit.

Nuclide	Half-life (s)	Specific activity (Bq/m^3)		Regulation limits (Bq/m^3)	Ratio	
		250 m^3	1000 m^3		250 m^3	1000 m^3
H 3	3.89E + 08	2.24E − 02	3.18E − 02	2E + 03	1.12E − 05	1.59E − 05
Be 7	4.61E + 06	3.32E − 01	4.42E − 01	1E + 03	3.32E − 04	4.42E − 04
C 11	1.22E + 03	6.04E + 02	8.47E + 02	2E + 04	3.02E − 02	4.24E − 02
N 13	5.98E + 02	1.37E + 03	1.87E + 03	6E + 03	2.28E − 01	3.11E − 01
O 15	1.22E + 02	6.16E + 02	8.48E + 02	3E + 03	2.05E − 01	2.83E − 01
Si 31	9.44E + 03	1.39E + 00	9.29E − 01	9E + 02	1.55E − 03	1.03E − 03
P 32	1.23E + 06	3.24E − 02	5.26E − 02	2E + 01	1.62E − 03	2.63E − 03
P 33	2.19E + 06	9.11E − 03	3.42E − 02	5E + 01	1.82E − 04	6.83E − 04
S 35	7.56E + 06	5.93E − 03	1.20E − 02	5E + 01	1.19E − 04	2.39E − 04
S 38	1.02E + 04	8.67E − 01	1.73E + 00	4E + 02	2.05E − 03	4.10E − 03
Cl 38	2.23E + 03	4.07E + 01	6.60E + 01	1E + 03	4.07E − 02	6.60E − 02
Cl 39	3.34E + 03	2.74E + 01	5.16E + 01	1E + 03	2.74E − 02	5.16E − 02
		Total sum			**5.37E − 01**	**7.63E − 01**

Table 3. Specific activity and ratios in two different volumes.

adults and 70-year commitment period for children (ICRP2001 [11]). Dose coefficients (committed effective dose per unit intake) are published in the summation of accumulated doses over the commitment period. Different dose coefficient values are specified for lung absorption type (ICRP, 2001) or lung inhalation class (U.S. Environmental Protection Agency, 1988 [12]), which relates to the solubility of inhaled material in lung fluid and clearance rate (fast, moderate, and slow) from the pulmonary region of the lung, respectively.

To calculate the radiation exposure of a personnel staying in an hour in an atmosphere with unrestricted release limit, first, calculate the isotope-dependent effective dose rate per Bq/m^3 by means of the unrestricted release concentration of radionuclide (C_i), which is given in table by the National Nuclear Safety Commission. For a member of the public, maximum annual source constraint of effective dose (300 μSv/year) is applied.

$$\frac{\text{Effective dose rate}}{\text{Specific air activation}} = \frac{\left(\frac{\mu\text{Sv}}{\text{h}}\right)}{\left(\frac{\text{Bq}}{\text{m}^3}\right)} = \frac{\frac{300}{365\times 24}}{C_i} \tag{2}$$

For conservative estimation, review of the unrestricted release limit concentration of radionuclide C_i is followed by checking the consistency of maximum annual dose rate. In this calculation, dose coefficient $h(g)_{\max}$ and a respiratory rate 1.2 m^3/h for average adult are used in addition to the tabulated radionuclide concentration C_i.

$$\frac{\text{Dose}}{\text{Year}} = \frac{24\text{h}}{\text{d}} \times \frac{365\text{d}}{\text{y}} \times \frac{1.2\text{m}^3}{\text{h}} \times C_i \times h(g)_{\max} \tag{3}$$

If this value is bigger than 300 μSv/year, unrestricted release limit C_i must be reduced for conservative evaluation. Thus, a new adjusted unrestricted release concentration of radionuclide (C_i^{new}) is calculated.

$$C_i^{\text{new}} = C_i^{\text{old}} \times \frac{\frac{300\mu\text{Sv}}{\text{year}}}{\frac{\text{Calculated dose}}{\text{year}}} \tag{4}$$

For each single radionuclide i, the rescaled unrestricted release limit concentration value is used to calculate the radiation exposure of a personnel who is staying in an air with a given radionuclide. The result is conservative and takes into account not only inhalation but also the dose taken up by immersion and ingestion. To calculate the committed accumulated dose, specific air activity calculated by the CINDER'90 code and conversion factors calculated above are used as a summation for all radionuclide.

$$\text{Total inhalation dose} = \sum_i (\text{intake}_i) \times h(g)_{\max}, \tag{5}$$

where (Intake$_i$) is inhalation intake of radionuclide i (Bq) and $h(g)_{\max}$ is inhalation dose coefficient for radionuclide i (Sv/Bq).

However, in this calculation, the ventilation-induced air exchange is not taken into account. In this study, the average stay length of the air molecules in the room is taken into account. Individual radionuclides, multiple sources of contamination, and different physical activity levels are accounted for by performing computations for separate contributions followed by summation. Inhalation dose coefficients have been tabulated for individual radionuclides in ICRP2001, which are consistent with the dosimetric model and organ weighting factors of ICRP1991 [13] and ICRP1995 [14], often referred to as ICRP-60 dosimetry and ICRP-72 dose coefficients.

Finally, radiological impact of medical staff that prepares and performs a patient treatment is estimated. In this study, it is assumed that a staff member spends 2 min treatment time and 28 min preparation time for a total 30 min exposure time per patient. Air ventilation rate is 20 min so that the staff member stays 18 min in the activated air with no ventilation. **Figure 11** shows the dose rate and accumulated dose for a normal single fraction of treatment scenario.

Dose rate decreases nonlinearly at the start of a treatment because of the high production of short half-life isotopes such as C-11, N-13, and O-15. The room is refreshed after 20 min by the air ventilation system. The early dose rate is very high but it is less than 0.11 μSv/h after 3 min of decay time. Thus, it is desirable for the staff to enter the room a couple minutes later after the end of a treatment irradiation. The total committed effective dose is about 0.021 μSv for one fraction of treatment. If 1000 patients are assumed to be treated in a year and 12 fractions per each patient, then a total of 12,000 fractions per year are irradiated, and the total effective dose is about 252 μSv/year which is approximately equal to the measured natural dose in the area.

4.2. Cooling water activation

Accelerators and their electromagnet components are cooled by using desalinated cooling water. If there is a non-negligible level of beam leakage in any part of the component, radioactive material may be generated in the cooling water passing around it. In general, accelerator parts and cooling water are replaced periodically. Replacement cycles vary depending on the type and nature of the component but mostly, it is extremely rare to continue using it for more than 10 years. In this section, for a conservative estimation, it is assumed that the cooling water has been used for 10 years.

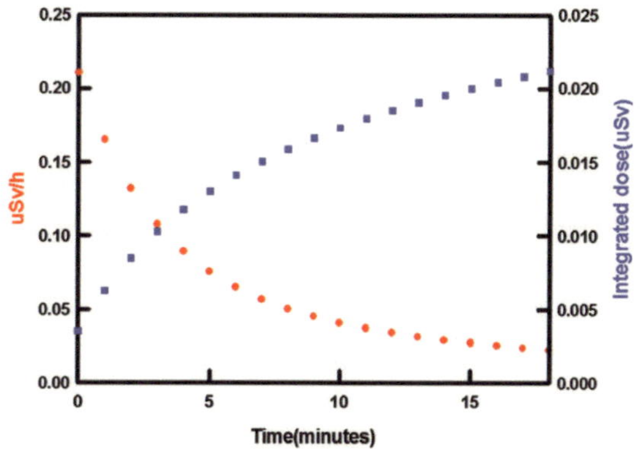

Figure 11. Effective dose rate and its accumulated dose in normal treatment, staying for 18 min in the room with no ventilation.

In the accelerator system, the region where the highest level activation of cooling water is frequently generated is mainly the region where the high-energy particles are lost or the region where the beam loss point and the flow of the cooling water are close to each other. Considering these two factors together, the high-energy beam loss area in an accelerator system is the magnetic extraction septum. Specifically, extraction magnetic septum is the highest energy beam (430 MeV/u) loss point. Thus, conservative estimation of all cooling water activation in an accelerator system can be expended based on the results of the extraction magnetic septum evaluation.

4.2.1. *Extraction magnetic septum in an accelerator system*

Septum makes the high-energy beam to escape to the left or continue to rotate inside the accelerator. The beam loss is about 5% (~0.5×10^8 pps). The lost beam interacts with the cooling water line made of copper to generate radioactive nuclides in the cooling water. The total amount of cooling water used in the system is about 30 tons, of which about 8 tons remains in the reservoir and the rest is running on the line. All the cooling water lines are designed in an independent way to cool one part of the system and return to the heat exchanger. In other words, cooling water that has cooled one part does not cool other parts before it reaches the heat exchanger. The total length of the cooling water line is about 90 m on average and the cooling water is designed to rotate about once a minute. The speed of the cooling water is 1.5 m/s. Thus, the exposure time of the cooling water (in the 0.7 m extraction magnetic septum) to the beam is about 0.5 s which is 1/120 (irrational number) of the actual beam use time. In this study, the computation is performed at about 1/100 (rational number) of the actual beam use time to obtain more conservative results.

Figure 12 shows an extraction magnetic septum. Bottom projection shows four square conductors with an edge length of 9.8 mm and four holes of 3 mm diameter each, which are used as water cycles. The septa have four parallel arranged square head. The length of a water flow is about 694 mm (middle).

4.2.2. *Simulation model 1 of the extraction magnetic septum*

As shown in **Figure 13**, the amount of cooling water in each line is 5.0 ml for a total of 20 ml in four lines. A beam loss on the front surface of a 70 cm long cylindrical copper rod is modeled by a radius of 3 mm copper rod. This radius corresponds to the thickness of the copper layer, which is located between the magnetic field in the septum and the water cycle. The copper rod is surrounded by concentric cylinder of radius 5.0 mm with cooling water over a length of 70 cm. The entire arrangement is surrounded by an iron cylinder with a height of 74 cm and a radius of 50 cm in order to take account of the back scattering of secondary particles in the volume of water. This generic model conservatively covers the radionuclide of the cooling water, which is running in the extraction magnetic septum because the total amount of water in this model is 35 ml.

4.2.3. *Simulation model 2 of the extraction magnetic septum*

Model 2 is the same as model 1, but replaces copper pipes with cooling water. This model is a conservative model for the case where the lost beam from the front of the magnetic septum

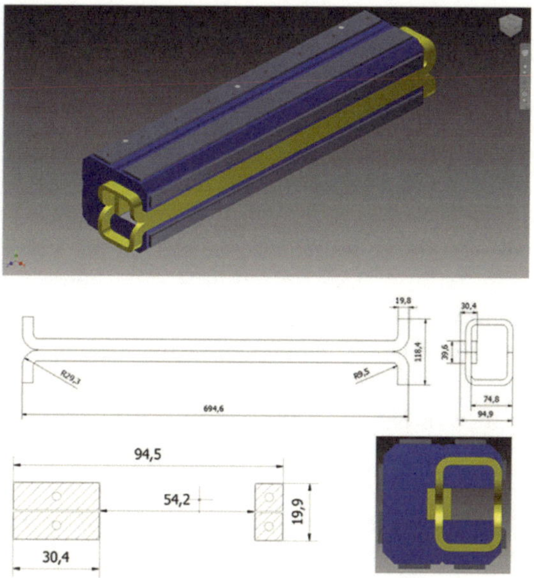

Figure 12. Water cooled extraction magnetic septum. Top: 3D drawing. Middle: cooling water line cross section drawings. Bottom: its projection.

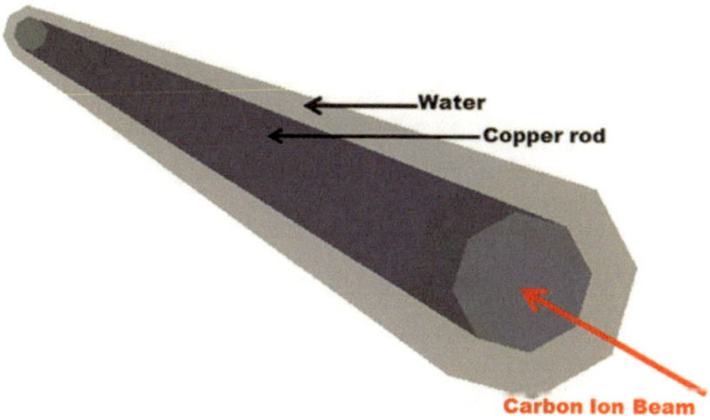

Figure 13. Schematic diagram for cooling water model 1. Outer iron cylinder is not shown here.

directly interacts with water. Thus, high-energy primary carbon particles interact directly with water, thereby maximizing the emission of cooling water. In particular, the lost beam becomes parallel to the cooling water flow, leading to the maximum activation of the cooling water.

4.2.4. Simulation model 3 of the cooling water

The final model calculates the cooling water emission in a cylinder with a height of 4 m and a radius of 1 m. The beam is incident on the front surface of the cylinder in a direction parallel to the cylinder axis. This model allows for the most conservative estimation because it calculates all possible interactions between cooling water and high-energy beams. The amount of total cooling water interacting with the beam is approximately 42% of the 30 tons of the total cooling water used in the system. It is highly conservative simulation because five times more amount of water is exposed to the lost beam than the amount of cooling water in the actual system. In this respect, model 3 is the most conservative model to explain all cases of the cooling water emissions.

4.2.5. Results and discussion

The decay times for prolonged release were set to 1 day, 1 week, and 1 month. For 1 day, decay just after 10 years normal operation, the results for specific activity and total activity for the three models are shown in **Table 4**. Specific activities were compared with regulatory limit values. The radioactive isotopes contributing to the activation of the cooling water are mainly H-3 and Be-7. Tritium has a half-life of about 12.4 years. That is, the cooling water used for 10 years did not saturate the tritium. However, Be-7 has short half-life of about 53 days. Therefore, although the effect of tritium is increasing when it is used for a longer period of time, the influence of Be-7 may be larger when it is used for a short time. The activation in the desalted water is relatively simple in this way. However, it is considered that the radioactive phenomenon of impurities which may be caused by the corrosion of the cooling water pipe is very complicated and it is impossible to carry out the detailed simulation here.

Decay time	**Nuclide**	**Regulatory limit (Bq/m^3)**	**Activity**		**Specific activity/ regulation limit**
			Total activity (Curies)	**Specific activity (Bq/m^3)**	
Model 1	H 3	2.00E + 07	6.86E−06	8.46E + 03	4.23E−04
	Be 7	2.00E + 07	1.77E−06	2.19E + 03	1.10E−04
	Be 10	6.00E + 05	1.34E−12	1.65E−03	2.75E−09
	C 11	3.00E + 07	5.36E−28	6.61E−19	2.20E−26
	C 14	1.00E + 06	1.71E−09	2.10E + 00	2.10E−06
	Sum				**5.35E−04**
Model 2	H 3	2.00E + 07	9.09E−05	1.12E + 05	5.60E−03
	Be 7	2.00E + 07	1.31E−04	1.61E + 05	8.05E−03
	Be 10	6.00E + 05	7.90E−11	9.75E−02	1.63E−07
	C 11	3.00E + 07	3.70E−26	4.56E−17	1.52E−24
	C 14	1.00E + 06	5.42E−08	6.68E + 01	6.68E−05
	Sum				**1.37E−02**
Model 3	H 3	2.00E + 07	6.64E−04	8.19E + 05	4.10E−02
	Be 7	2.00E + 07	3.79E−04	4.67E + 05	2.34E−02
	Be 10	6.00E + 05	2.89E−10	3.57E−01	5.95E−07
	C 11	3.00E + 07	1.28E−25	1.58E−16	5.27E−24
	C 14	1.00E + 06	2.96E−07	3.65E + 02	3.65E−04
	Sum				**6.47E−02**

Table 4. Comparison of 1 day decay results just after 10 years normal operation for model 1, model 2, and model 3.

4.3. Ground water and underground soil activation

4.3.1. Simulation model

Some of the primary and secondary beams are lost to the bottom of the building during beam transport and interact with the floor concrete, underground soil, or groundwater to generate radioactive isotopes. Detailed evaluation is necessary because radioactive nuclides in underground soil or groundwater are directly related to environmental pollution. Underground soil or groundwater is located 2.5 m below the floor concrete of the facility.

A simple cylindrical generic model is created for convenient calculation. In order to mimic the distance of 120 cm between concrete floor and the beam loss point, a cylinder-shaped concrete model of inner radius of 120 cm with cylindrical iron target of length of 50 cm and a diameter of 5 cm at the center is devised, and a maximum energy carbon beam is incident in the axial direction. The distance, 120 cm, between the target and the concrete is the minimum distance between the beam line and the ground floor. Thus, the model is conservative. The thickness of the concrete is 2.5 m, which is the same as the floor concrete thickness of the facility. Underground soil or groundwater is modeled about 10 cm thick on the outside of the concrete.

The environmental pollution of the facility is evaluated by the radioactivity of about 10 cm thick soil or water. This model can also be used for the analysis of the floor concrete activation. Fifty years operation with maximum energy is assumed for a conservative evaluation of the activation of underground soil or groundwater. Six collapse scenarios apply from 1 min to 1 year after 50 years of normal operation, required by the radiation management regulation. **Figure 14** shows underground model. Blue area (or outside area) represents groundwater or underground soil.

4.3.2. Results and discussion for groundwater

The result of 1 min decay time is shown in **Table 5**. Radionuclides with relatively long half-lives such as H-3, Be-7, Be-10, and C-14 are observed, which is similar to the cooling water activation. However, the contribution of H-3 is about 76% or more, and the remaining isotopes contributions are only about 24%. The results suggest that the lost beam has little effect on the groundwater below the ground floor concrete because of the thickness of the floor concrete. The total sum of the isotope concentration ratios is only about 10^{-5} at the end of beam

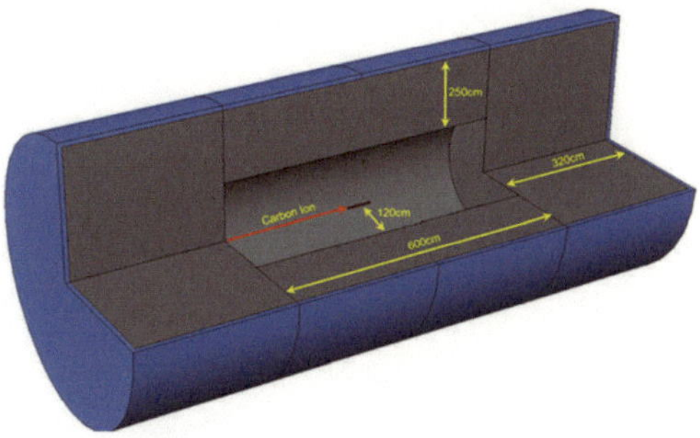

Figure 14. Generic geometry model for the floor concrete, underground soil, or groundwater simulation.

irradiation, equivalent to 50 years normal operation, so that there is no harmful effect of radioactivity to the groundwater.

4.3.3. Results and discussion for underground soil

Unlike the activation of groundwater, the activation estimation of underground soil can take a long time. In other words, it is common that the dismantling of a facility takes about 1 year after complete closing of the facility. Thus, the activation analysis of the underground soil is performed, based on 1 month and 1 year of decay time. **Table 6** shows the underground soil activity concentrations (at 1 year decay time) generated after 50 years of normal operation with maximum energy. About 10 different isotopes are contributed and K-40 is conspicuous and occupies 99%.

Decay time	Nuclide	Specific activity (Bq/m^3)	Regulation limit (Bq/m^3)	Specific activity/regulation limit
1 min	H 3	4.53E + 02	2.0E + 07	2.26E − 05
	Be 7	1.63E + 01	2.0E + 07	8.13E − 07
	Be 10	2.47E−04	6.0E + 05	4.12E − 10
	C 11	1.92E + 02	3.0E + 07	6.41E − 06
	C 14	6.53E − 02	1.0E + 06	6.53E − 08
	N 13	1.39E + 02	—	—
	O 14	1.26E + 02	—	—
	O 15	1.45E + 03	—	—
		Sum		**2.99E − 05**

Table 5. Ground water activation.

Decay time	Nuclide	Specific activity (Bq/g)	Regulation limit (Bq/g)	Specific activity/regulation limit
1 year	K 40	2.76E − 01	1.00E + 02	2.76E − 03
	Na 22	8.00E − 05	1.00E + 01	8.00E − 06
	Mn 54	3.71E − 05	1.00E + 01	3.71E − 06
	Sc 46	6.56E − 07	1.00E + 01	6.56E − 08
	Fe 55	9.23E − 05	1.00E + 04	9.23E − 09
	Fe 59	6.00E − 09	1.00E + 01	6.00E − 10
	V 49	5.09E − 06	1.00E + 04	5.09E − 10
	Al 26	3.37E − 09	1.00E + 01	3.37E − 10
	H 3	3.35E − 04	1.00E + 06	3.35E − 10
	Be 7	4.66E − 08	1.00E + 03	4.66E − 11
		Sum		**2.78E − 03**

Table 6. Radiological impact of underground soil after 50 years normal operation.

4.4. Accelerator materials activation

4.4.1. Simulation model

The accelerator parts are very likely to be activated because the lost beam in the accelerator interacts with the periphery while transported to the irradiation room. In this section, the maximum doses of the highest beam loss parts in the accelerator, magnetic extraction septum and beam dump, are analyzed. It is desirable to understand precisely the physical and theoretical activation of the whole accelerator system with a simpler and conservative model. The reason why magnetic extraction septum is chosen is that most of the highest energy beam is lost at this part and irradiated to the copper coil consisting of the septum. Meanwhile the beam that survives interacts with the iron york shielding it.

In this model, a big thicker septum structure, compared to the one used in the calculation of the cooling water, is chosen for conservative estimation. A copper wire with a length of about 102.5 cm, a cross-sectional area of 9.8 mm × 9.8 mm with a radius of 3 mm cylinder with cooling water flowing is embedded in the outer iron york. Because the amount of radioactive nuclides is proportional to the total mass or volume of the material, a simpler model with a mass or volume of the same material is used to calculate the activation of the materials. Thus, the copper cylinder of the same volume (1 m in length and 1.5 cm in radius) is modeled inside the concrete structure. The outside concrete structure is used to simultaneously calculate the activation of the concrete floor placed along the beam line. **Figure 15** shows the thick magnetic extraction septum with eight cooling water lines of a length 102.5 cm and diameter 1.5 cm.

The distance between the concrete and the target material was modeled to be about 120 cm apart along the actual accelerator beam line array for a conservative estimation, which is similar to what is shown in **Figure 14**, excluding the groundwater part. A maximum energy of 430 MeV beam was irradiated to the copper rod of length 1 m and radius of 1.5 cm for 20,000 h. Six decay times were applied from 1 min to 1 year. A total of 20,000 h are assumed to be a life time of extraction magnetic septum.

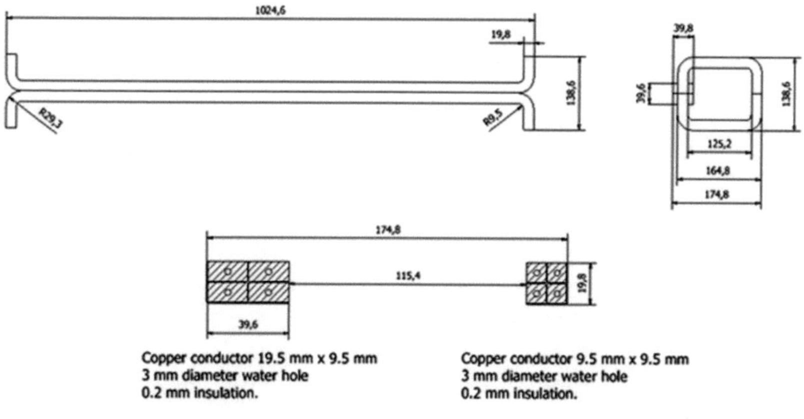

Figure 15. Thick magnetic extraction septum.

4.4.2. Results and discussion

The sum of the ratios of specific activity concentration to regulation limit values were 290, 242, and 83.2 for decay times of 1 week, 1 month, and 1 year, respectively. Notice that the sum of the ratios is greater than unity. Accordingly radioactive waste must be treated in accordance with radiation regulations. Important isotopes that cause most of the radioactivity problems after 1 year of decay time were slightly different from those of the 1 h decay time. In particular, Co-60 accounted for more than 50%. In addition, Mn-54, Na-22, Zn-65, Co-56, Co-57, and Co-58 contributed greatly.

F4 tally was used to calculate the average effective dose rate of air around copper and F5 tally was used to estimate the effective dose rate at the center point of the copper wire. The results were 3.42×10^{-3}, 2.04×10^{-3}, 1.42×10^{-3}, 1.21×10^{-3}, 9.67×10^{-4}, and 3.82×10^{-4} Sv/h for decay times of 1 min, 1 h, 1 day, 1 week, 1 month, and 1 year, respectively. The results suggest that, if possible, repair must be done after 1 month of decay time. If emergency repair needs, necessary radiation protection equipment must be used.

Gamma rays effect immediately after beam irradiations were about 3 μSv/h at a distance of about 2 m from the septum. In normal operation, the average prompt and delay gamma rays effect were about 0.5 μSv/h. After 1 h decay time, the effect was reduced by a factor of 10. Thus, an appropriate action must be taken for any emergency work. **Figure 16** shows the surrounding area effective dose rate immediately after an hour irradiation to the copper target with maximum intensity of 1×10^9 pps as a function of decay time for just after beam stop, 1 min, 1 h, 1 day, and 1 week.

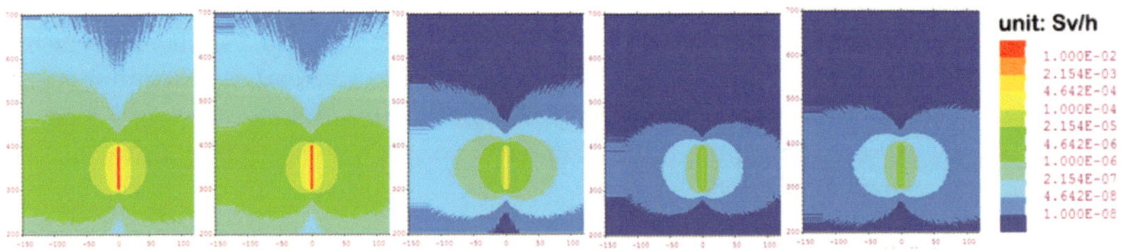

Figure 16. Effective dose (due to gamma rays) of surrounding area after an hour irradiation to the copper target with maximum intensity of 1×10^9 pps and maximum energy of 430 MeV/u as a function of decay time. From left to right, just after beam stop, 1 min, 1 h, 1 day, 1 week, respectively. Copper rod (with the maximum dose profile) is shown in the central area.

5. Conclusions

Analysis of nuclear safety is an essential requirement for optimizing the design of high-energy accelerator facilities and for preparing safety analysis report. Many novel generic models for nuclear safety analysis were introduced by using an example accelerator facility under development in Korea. Well-known MCNPX2.7.0 and CINDER'90 codes were used for nuclear transportation and transmutation simulations. Shielding simulations were performed according to the NCRP 147 recommendations with maximum occupancy factor and maximum workload for conservative estimation. The results of skyshine and groundshine analysis showed the possibility of the internal radiation affecting to the surrounding environment.

In the field of radioactivity, air activation, cooling water activation, activation of groundwater and underground soil, and their effects on the external environments, medical staffs, and general public were analyzed. The activation phenomena of the accelerator materials were analyzed through the activation analysis of copper which is a typical material used for the accelerator. Access availability to the accelerator hall for emergency maintenance was analyzed by the estimation of effective dose rate surrounding the extraction magnetic septum in the accelerator. This new design via simulations brought many useful outputs with respect to the cost, flexibility, and simplicity of facility construction. Specifically, the decision of appropriate wall thickness and ventilation time etc. via simulations is an essential for cost effective construction of the facility.

Acknowledgements

The project is supported by the Institute for Modelling and Simulation Convergence (IM&SC).

A. Assessment of unrestricted release limit concentration for the unspecified isotopes

For some type of radioactive nuclides, unrestricted release limit is not specified by the National Nuclear Safety Commission. In such a case, the smallest value in the similar mass nuclide group is chosen but the result may be less realistic. More realistic unrestricted concentration can be estimated by using the known release limit value of radioisotope with the following steps.

Step 1: select a known radioactive isotope whose chemical properties are the similar to those of unknown radioisotope. For example, to calculate the unrestricted release limits for Cl-34 m and S-38, the radioisotope with a similar biological half-life must be sought. The biological half-life is:

$$\frac{1}{T_{1/2\text{effective}}} = \frac{1}{T_{1/2\text{physical}}} + \frac{1}{T_{1/2\text{biological}}} \tag{A1}$$

Step 2: among the isotopes listed in the regulation table, select the isotope whose value is known to be similar in terms of the radiation emission. At this time, the proof of similarity is determined by the Q value of the decay process and by the average energy of electrons or positrons in beta decay.

Step 3: use the following equation.

$$C_i(Iso_2) = C_i(Iso_1) \times \frac{1}{\frac{T_{1/2\text{effective}}(Iso_2)}{T_{1/2\text{effective}}(Iso_1)} \times \max\left[\frac{Q_{\text{value}}(Iso_2)}{Q_{\text{value}}(Iso_1)}, \frac{E_{\text{beta}}(Iso_2)}{E_{\text{beta}}(Iso_1)}\right]} \tag{A2}$$

	$T_{1/2bio}$	$T_{1/2phy}$	$T_{1/2eff}$	Q_{val}	E_{beta}	$\frac{T_{1/2eff}(Iso_2)}{T_{1/2eff}(Iso_1)}$	$\frac{Q_{val}(Iso_2)}{Q_{val}(Iso_1)}$	$\frac{E_b(Iso_2)}{E_b(Iso_1)}$	C_i
Cl-36	29 days	3×10^5 years	29 days	708 keV	247 keV	7.66×10^{-4}	7.76	1.69	0.1 Bq/m^3
Cl-34 m	29 days	32 min	32 min	5491.3 keV	418 keV				16.83 Bq/m$_3$
S-38	90 days	172 min	172 min	2937 keV	490 keV	2.70×10^{-3}	17.5	10	423.28 Bq/m^3
S-35	90 days	87.3 days	44.3 days	167.14 keV	49 keV				20 Bq/m^3

Table 7. Summary of concentration assessment.

Cl-36 is reference isotope for Cl-34 m and S-35 for S-38. The assessment is summarized in the following **Table 7**.

Author details

Oyeon Kum[1,2*]

*Address all correspondence to: okum@uswa.ac

1 Kyungpook National University, Bukgu, Daegu, Korea

2 University of Southwest America, Los Angeles, CA, USA

References

[1] AERB. Safety Guidelines on Accelerators. Atomic Energy Regulatory Board; Mumbai, India, 2005

[2] Yim H, An D-H, Hahn G, Park C, Kim G-B. Design of the KHIMA synchrotron. Journal of the Korean Physical Society. 2015;**67**:1364-1367. DOI: 10.3938/jkps.67.1364

[3] Pelowitz DB, editor. MCNPX User's Manual Version 2.7.0. LA-CP-11-00438. April 2011

[4] Wilson WB, Cowell ST, England TR, Hayes AC, Moller PA. Manual for CINDER'90 Version 07.4 Codes and Data. LA-UR-07-8412. Los Alamos National Laboratory, New Mexico, USA; 2008

[5] NCRP. Structural Shielding Design for Medical Use of X-rays Imaging Facilities. NCRP Report No. 147. 2004

[6] FLUKA. http://www.fluka.org/ [Accessed Jul 2017]

[7] Kum O, Heo S-U, Choi S-H, Song Y, Cho S-H. Radiation protection study for the first Korea heavy-ion medical accelerator facility. Nuclear Techniques. 2015;**192**:208-214. DOI: 10.13182/NT14-121

[8] ICRU. Tissue Substitutes in Radiation Dosimetry and Measurement International Commission on Radiation Units and Measurements Report 44. Bethesda, MD, USA; 1988

[9] ICRP. Human Respiratory Tract Model for Radiological Protection. ICRP Publication 66; 1994

[10] ICRP. Limits for Intakes of Radionuclides by Workers. ICRP Publication 30; 1979

[11] ICRP. Doses to the Embryo and Fetus from Intakes of Radionuclides by the Mother. ICRP Publication 88; 2001

[12] Eckerman KF, Wolbarst AB, Richardson ACB. Limiting Values of Radionuclide Intake and Air Concentration and Dose Conversion Factors for Inhalation, Submersion, and Ingestion. Washington, DC: Office of Radiation Programs U.S. Environmental Protection Agency; 1988

[13] ICRP. Recommendations of the International Commission on Radiological Protection. ICRP Publication 60; 1991

[14] ICRP. Age-dependent Doses to the Members of the Public from Intake of Radionuclides—Part 5 Compilation of Ingestion and Inhalation Coefficients. ICRP Publication 72; 1995

Chapter 6

Radiation Safety Aspects of Linac Operation with Bremsstrahlung Converters

Matthew Hodges and Alexander Barzilov

Additional information is available at the end of the chapter

http://dx.doi.org/10.5772/intechopen.71317

Abstract

This chapter provides a discussion of radiation safety aspects of operation of electron linear accelerators equipped with bremsstrahlung converters. Electron accelerators with 3, 6, 9 and 15 MeV electron beams are discussed. High-energy photon and photoneutron production during linac operation was analyzed using Monte Carlo methods. Radiation dose rates for different configurations of linacs were evaluated and compared with experimental results.

Keywords: linac, bremsstrahlung, photoneutrons, MCNPX, dose rate, radiation safety

1. Introduction

A linear accelerator (linac) is a system that increases kinetic energy of charged particles using oscillating electric potentials along the line of a beam of the particles (e.g., electrons, protons, ions). Within a few meters, it is possible for 10 keV electrons to be accelerated by the RF linac to up to 20 MeV [1]. Accelerators including linacs have found use in a variety of applications including radiotherapy [2, 3], physics [4, 5], isotope production [6], cargo inspection [7, 8], and non-destructive assay [9, 10].

In inspection systems and non-destructive assay applications, electron linacs are used to generate high energy photons that can penetrate objects under scrutiny. The accelerated electrons are bombarded onto a target composed of high-Z material. The incident electrons are deflected by electric field of the electron cloud of atomic nuclei of the target material, losing kinetic energy that is converted into the bremsstrahlung (or *braking radiation*). The bremsstrahlung photons produced by a linac are characterized by their energy distribution (the quantity of photons produced at specific energies). Linacs generate photons which have endpoint energy equal to the maximum energy of the electrons in the beam impinging on the target. In a 10 MeV linac, the bremsstrahlung has continuous spectrum from 0 up to 10 MeV photons. These photons are typically focused into a desired beam shape by the use of collimators (e.g., fan beams, conical beams). The beam of collimated photons is then used for a variety of applications including imaging, radiotherapy, the production of medical isotopes, or to perform an active assay of unknown materials. The photon flux is the number of photons passing through a defined area (e.g., 1 cm^2) per unit of time.

If the energy of an incident photon is greater than that of the neutron binding energy of material it interacts with, a neutron can be produced through the (γ,n) reaction. At photon energies greater than 10 MeV, the (γ,n) reaction will take place within materials that commonly compose the accelerator facility structures [11].

Radiation safety aspects are very important in operation of linacs with bremsstrahlung converters. To describe the effects of high-energy photons and photoneutrons produced by the linac on materials and personnel, it is necessary to quantify the amount of energy deposited by radiation when it interacts with matter. The term *dose* describes the amount of energy deposited by radiation within the material, while the *biological dose* describes the energy deposited in a living tissue. The biological dose equivalent is the dose multiplied by a quality factor Q used to express the biological damage variation between the different radiation types. The quality factor has the following values: $Q = 1$ for x-rays, gamma rays, or beta particles; $Q = 20$ for alpha particles and heavy ions including fission fragments; $Q = 10$ for neutrons and for high-energy protons [12].

The international system (SI) unit of dose is the *Gray* (Gy) and is equal to the absorption of 1 joule of energy by 1 kg of material. The *Rad* is equal to 1/100 Gy. The *Rem* is equal to the product of the absorbed dose and the quality factor. The terms *dose rate* and *biological dose rate equivalent* are used to describe the respective doses received per unit time. Once radiation fluxes in an environment have been characterized, the dose rates can be determined through the use of the energy-dependent flux-to-dose conversion factors for a specific radiation type. Several flux-to-dose conversion factors have been established (i.e., ANSI/ANS 6.1.1-1997 [13]).

Two Varian linacs—the M6 and K15 models equipped with bremsstrahlung converters—with 3, 6, 9 and 15 MeV electron beams were studied for operation at the shielded building. Radiation doses due to the linac operation in different configurations were analyzed using Monte Carlo modeling and experimental measurements.

2. Characterization of radiation generated by linac

It is important to understand the radiation generated during the linac operation and how that radiation is transported throughout the building in order to determine the location specific doses. Estimating the location specific dose rates within the building allows for verification of building safety measures, as well as help to evaluate expected dose rates at different distances from the linac structure which is vital for future research projects which may involve the irradiation of different sample materials.

Computational modeling is widely used to study behavior of complex systems. Models typically use numerous variables that characterize the system being studied. Simulation is performed by the adjustment of these variables and the subsequent observation of the outcome of the system. Computational modeling is often used as a first step in providing an estimation of parameters for a proposed experiment. When possible, model results should be validated against experimental measurement in order to determine the accuracy of simulations. Computational modeling is a valuable tool that allows for studying the effects of changing experimental parameters prior actually performing these experiments or designing radiation facilities, and for estimating their safety characteristics.

The Monte Carlo methods form a broad class of stochastic algorithms that proved successful in a variety of disciplines including genetics [14], space physics [15] and economics [16]. While problems might be solvable using deterministic methods, Monte Carlo methods utilize repetition of random sampling to arrive at a numerical result. With respect to nuclear science and radiation transport, Monte Carlo codes are used to track particle interactions with matter over a wide range of energies in a complex geometry [17, 18]. The computational modeling in this study is based on the Monte Carlo technique using the general purpose Monte Carlo N-Particle Transport software suite developed by Los Alamos National Laboratory—MCNP5 [19] and MCNPX [20] codes.

2.1. Monte Carlo model of linac

The accelerator facility at University of Nevada, Las Vegas was used as a representative building to study radiation safety aspects of operation of M6 and K15 linacs. The facility consists of an entry room, a shielding maze, and an accelerator bay. The control room is located south of the facility building itself, and is home to the operating controls of the linac as well as radiation detection equipment.

The ceiling in the facility is 20 cm thick concrete, with the walls and floors being 15 cm thick; also composed of concrete. The entry room is a large open space that measures approximately 11 m by 10 m. The shielding maze is formed by two walls of concrete bricks that serve to minimize the radiation doses in the entry way that are due to radiation emitted from the linac in the accelerator bay. The southern shield maze wall is 4 m long, 87 cm thick and the north wall is 7.5 m long and 117 cm thick. Both shielding walls are 2.5 m tall and extend almost completely to the ceiling. The scheme of the facility is shown in **Figure 1**.

The material compositions of the internal facility structures were taken from reference [21]. The M6 linac and K15 linac were modeled using different MCNP codes. The model of the M6 linac was run using MCNP5 with the ENDF/B-VII cross sections (denoted by the ".70c" identifier in MCNP5) at room temperature. The model of the K15 linac used the MCNPX code because it could not be run in MCNP5 due its lacking of photonuclear physics and the proper cross section library.

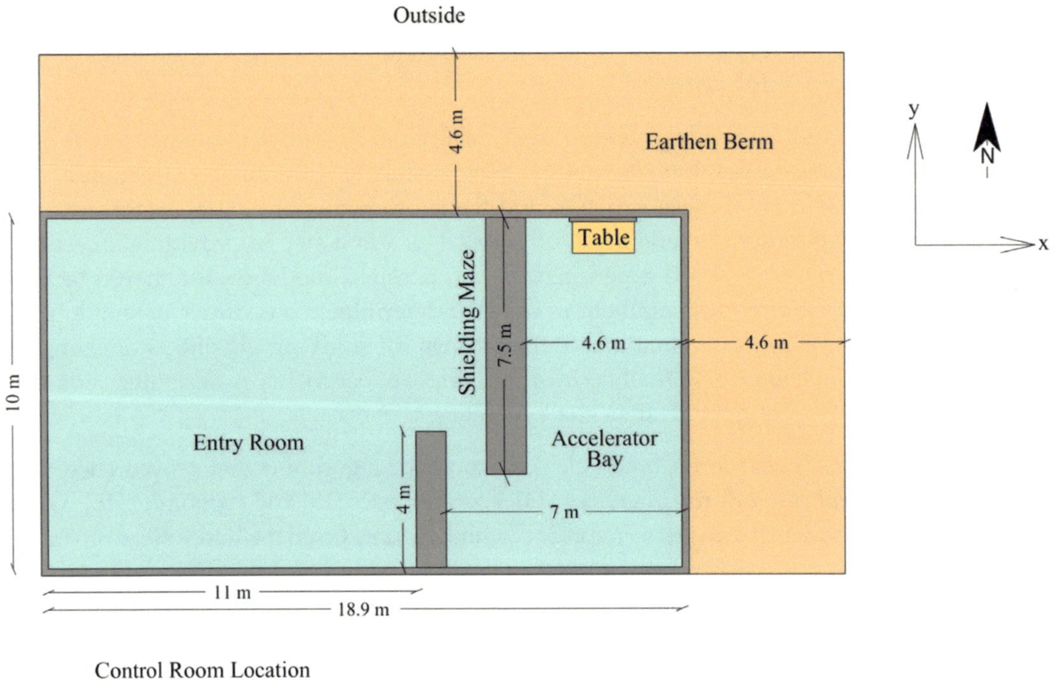

Figure 1. Accelerator facility layout.

2.2. Linac source definition

The computational models of the M6 and K15 linacs start with the simulation of an electron beam impinging upon the linac target head. The target head geometries and comprising material differ between the two linacs, and were each modeled according to the Varian specifications. In order for a computational model to produce precise results, it is necessary to accurately describe the source term. The source definition (SDEF) card is used to define the particle transport required within the MCNP model. An example of the MCNP5 input to define the electron source term is shown below:

```
sdef x d1 y 0 z d2 erg = d3 par 3 vec 0 1 0 dir 1.
sp1 −41 0.065 0.
sp2 −41 0.065 0.
sp3 −41 0.120 6.
```

The "mode p e" card was used to include photon and electron transport in the model. The "phys:p" card was used to include the production of bremsstrahlung by electrons. The source was defined to be a 1.3 mm electron pencil beam traveling in the *y*-direction. The *x* and *z* directions of the electron beam followed distributions *d1* and *d2* to use the built-in Gaussian probability (denoted by −41) for spatial coordinates extending 0.65 mm in both directions. The *erg* value specified the energy of the electrons in the beam using distribution *d3* to set the

Gaussian-type function centered on 6 MeV. The *par* value of the source card was set to 3 to specify an electron source and the *vec* and *dir* values were set to <0 1 0 > and 1 respectively, to specify the direction of the electron beam along the *y*-axis. The source term was checked to ensure the Gaussian nature of both the spatial coordinates and energy distribution (see **Figure 2** that shows 3 and 6 MeV cases for the M6 linac).

The K15 linac's source term was defined similarly, except the average energy in the *sp3* was set as 15 MeV (or 9 MeV in the low energy mode). Furthermore, neutrons were added to the "mode" card. The "phys:p" card default values must be modified to account for photonuclear production in the model. The *ispn* value (4th entry) on the "phys:p" card must be changed from 0 (default) to either −1 (the analog photonuclear particle production) or 1 (the biased photonuclear particle production). Additionally, the *fism* value (7th entry) of the "phys:p" card was set to 1 to enable the LLNL fission model (as opposed to the default ACE model). The LLNL model was used because the ACE model does not account for prompt photofission gamma rays [22].

2.3. Computational determination of radiation flux

To evaluate the bremsstrahlung photons produced within the respective linac target heads, it was necessary to determine their angular distribution and energy spectra. Thin (0.01 cm), concentric ring surfaces were set in the model in 1 cm behind the linac target head. F4 tallies (track length estimates of the cell flux) were set within each ring surface allowing for the determination of the x-ray flux at 10° increments off the centerline (see **Figure 3**). Two hundred equally spaced energy bins were used at each tally in order to determine the energy distribution of photons. The relative error associated with each bin in an MCNP tally (corresponding to one standard deviation) is given by the inverse square root of the number of source particles contributing to that tally. The MCNP suggestion is below 10% error for F4 tallies.

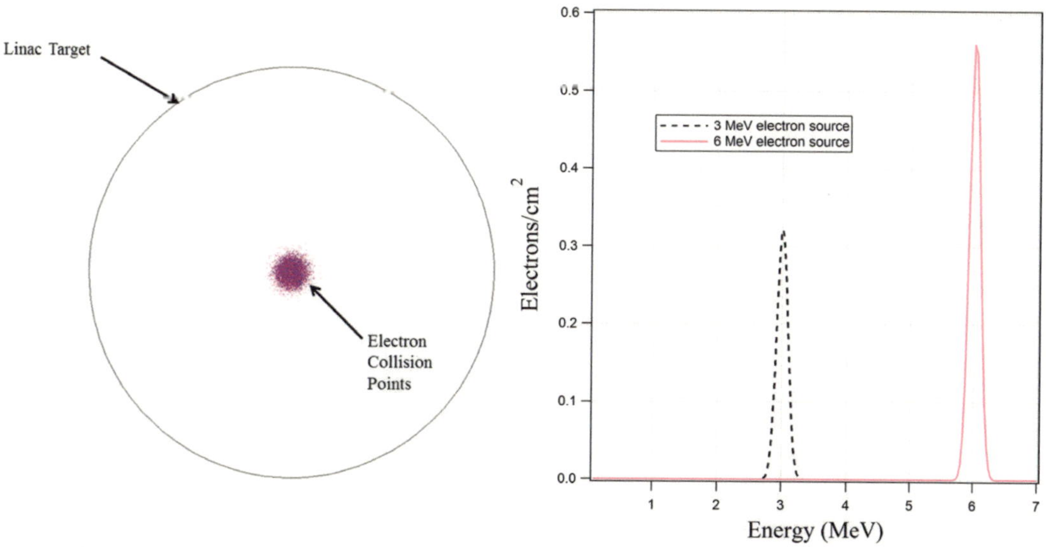

Figure 2. Spatial distribution (left) and energy distribution (right) of electrons on the M6 linac target.

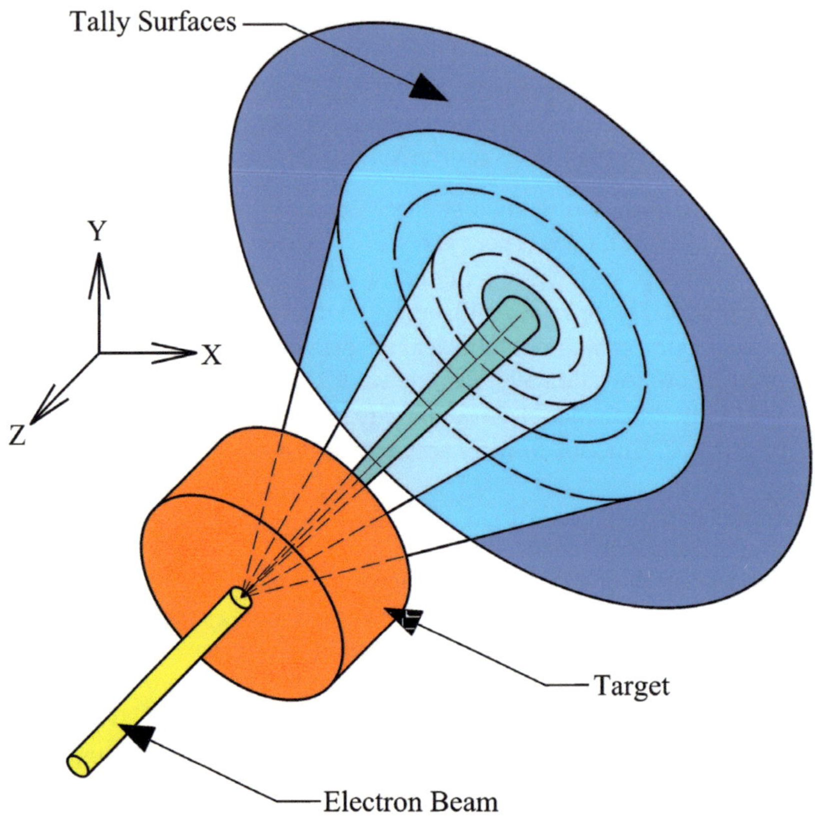

Figure 3. Schematic of tallying of bremsstrahlung angular distribution.

A mesh tally was used to determine the x-ray fluxes and dose rates due to operation of the M6 linac within the accelerator facility at the height of the fan beam (1.2 m above the floor). The FMESH card was used to determine the photon fluxes at 7.54 cm intervals in the x and y directions throughout the building. The dose energy (DE) and dose function (DF) cards were used in the MCNP5 model to convert the computed photon fluxes into dose rates by using the ANSI/ANS 6.1.1-1977 photon flux-to-dose rate conversion factor [13]. In addition to the FMESH tally, F5 tallies (flux estimators at a point) surrounded by the *dxtran* spheres were used to determine the dose rates at specific points within the facility. Moreover, the dose rate was measured by an ion chamber intrinsic to the M6 linac.

A mesh tally was similarly used to determine the photon and neutron dose rates within the accelerator facility for operation of the K15 linac, but the mesh tally syntax used in MCNPX differs from that used in MCNP5. In MCNPX, the TMESH tally with RMESH (denoting a rectangular mesh) was used to determine the dose rates throughout the building at the same spatial intervals as used in the M6 linac model. The dose rate conversions are handled in MCNPX within the RMESH by using the keyword DOSE and specifying the *ic* value as 20 (corresponding to the ANSI/ANS 6.1.1-1977 flux-to-dose factors) for both photons and neutrons.

The "NPS" card was used to set the particle history cutoff. Once the number of simulated particle histories exceeds the number specified by the NPS card, the MCNP model stops running and generates an output file, from which the results can be analyzed. Some models can take several days or weeks to complete depending on a computer, the intricacies of the model's physics, the cutoff value, and the type and energy of particles in the simulation. Charged particles (i.e., electrons) have large numbers of interactions due to the long-range Coulomb force whereas neutral particle interactions are defined by infrequent isolated collisions [23]. As such, simulations involving charged particle transport take longer to complete than those without a charge.

2.4. Determination of electron current on linac targets

The MCNP tally results are normalized per a starting particle. As the MCNP models of linacs in this study began with the simulation of electron transport, it was necessary to determine the number of electrons per second in the beam impinging the target in order to acquire quantifiable values for photon flux and dose rates. The M6 and K15 linacs utilize pulsed electron bunches to produce the bremsstrahlung radiation. As the electron current is not constant, it is required to determine the DC averaged current for each linac.

The voltage of a single pulse of the electron beam on the M6 linac target head was measured using a Teledyne Lecroy oscilloscope. The single pulse voltage waveform was converted to single pulse current by dividing by the resistance (50 Ω), determining the total area under the curve and multiplying by the frequency (156.555 Hz) to obtain a total DC averaged electron current of 3.4×10^{14} electrons per second. The electron current values for the K15 linac were provided by Varian.

3. Radiation fluxes and dose rates during linac operation

3.1. Bremsstrahlung spectra

The spectra of x-rays generated in the bremsstrahlung converter within the M6 linac were determined for the low (3 MeV electron beam) and high (6 MeV electron beam) energy operation modes with the results shown in **Figure 4**. The largest photon fluxes occur within 10° of the central axis of the linac target and decrease with increasing the angle. The trend of decreasing flux with increasing energy is apparent in both of the computed spectra. There is an order of magnitude difference between the fluxes at each angular interval between the two M6 operation modes. This is due to the greater likelihood that higher energy electrons will produce bremsstrahlung radiation with higher energy within the linac target. The error associated within each energy bin in the bremsstrahlung spectra are less than the MCNP recommended value of 10% for F4 tallies.

The radial variation of the photon flux within the conical segments was tested. Thin cylindrical cells were placed radially at 30° intervals throughout the conical segment. F4 tallies were set in each of these cells with the results showing that the flux at each radial location had similar

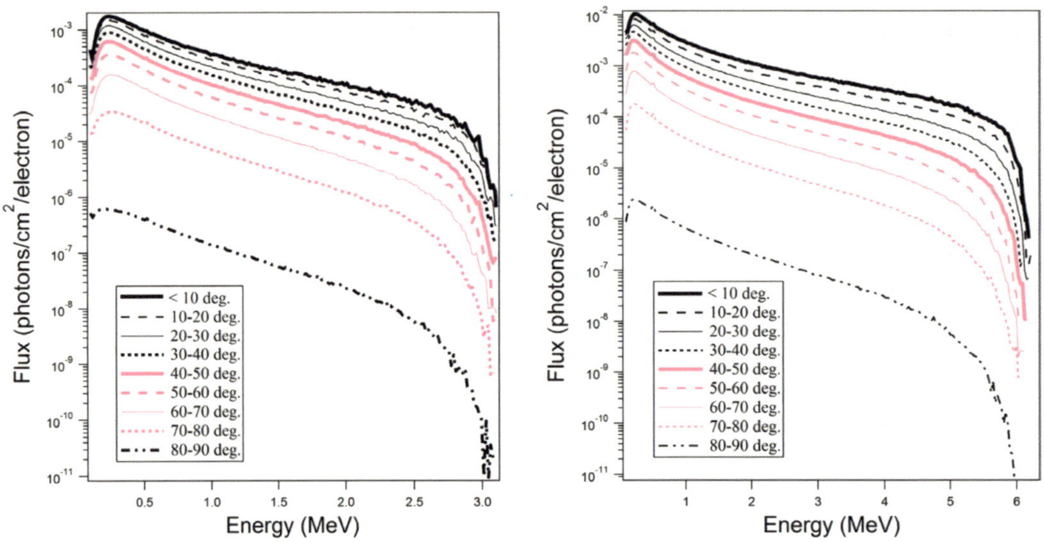

Figure 4. M6 linac: bremsstrahlung photon spectra, 3 MeV (left) and 6 MeV (right) incident electrons.

spectral distributions. Thus, the x-ray source exhibited the radial symmetry within conical segments. As the computational results for the flux are normalized to one starting particle (electron), they must be multiplied by the number of particles (described previously) in order to evaluate the photon flux generated by the target.

The bremsstrahlung spectra for the K15 linac were computationally determined for the low (9 MeV) and high (15 MeV) energy operation modes with the results shown in **Figure 5**. A total

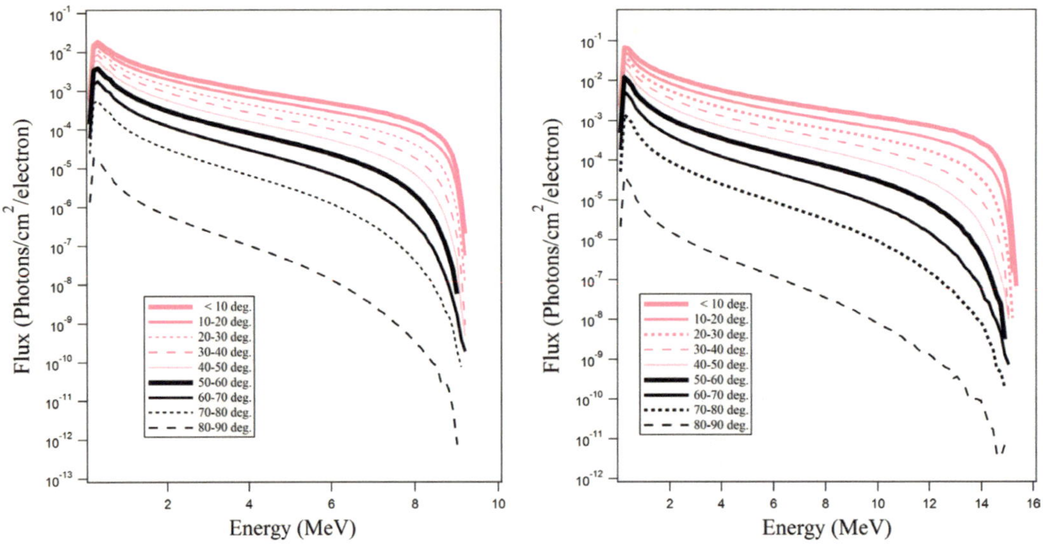

Figure 5. K15 linac: bremsstrahlung photon spectra, 9 MeV (left) and 15 MeV (right) incident electrons.

of 500 million particle histories were used in the simulation to ensure that the error associated with each of the 100 equally spaced energy bins was below the 10% recommended by MCNP for F4 tallies. As was the case with the M6 linac, the largest photon fluxes occur within 10° of the center line of the linac target and decrease with increasing outward angle. In both operating modes, at energy around 1 MeV, the photon fluxes are generated nearly three orders of magnitude larger within an angle of 10°, than they are at angles greater than 80°. This flux ratio increases to five orders of magnitude for 8 MeV photons. There is about an order of magnitude difference between the fluxes at each angular interval between the two K15 linac operation modes.

3.2. Electron energy cutoff

The MCNP5 models used to determine the M6 photon spectra were run using an NPS value of 10^8 starting electrons in order to minimize the error in each energy bin. Difficulty exists in determining the proper balance between minimizing computational time while maintaining satisfactory results. In order to reduce the computational time in this study, the suitability of using energy cutoff cards was investigated. Care must be taken when using energy cutoff cards as their use affects the underlying physics, resulting in the halting of particle interactions occurring under this energy threshold. In some instances, this may remove certain reactions from happening in the model or may modify results incorrectly beyond that which was originally determined. Results obtained from models using energy cutoffs should be compared against those without using energy cutoffs in order to make an accurate assessment as to whether their use is acceptable. The M6 bremsstrahlung spectra were determined using electron energy cutoffs of 0.001 (default), 0.01, 0.1 and 1.0 MeV with an NPS of 10^7 electrons. The results of the photon fluxes using the default and 1.0 MeV energy cutoffs for the first four energy bins are shown in **Table 1**.

The results showed that using the electron cutoff energy of 1.0 MeV in the MCNP5 model provided in photon flux values of above 70% similarity to the original results at all angle intervals within the first energy bin (below 0.1 MeV). In the second energy bin (between 0.1 and 0.16337 MeV), the average similarity rose to around 85% and by the third bin 90%. Above the third energy bin, average similarities between the two models rose to 95%. It was found that as the energy increased, so too did the similarity between model results for photon fluxes. Above the 1 MeV electron cutoff, the photon fluxes were identical. The computational time (rounded to the nearest minute) for running the MCNP5 model with the default energy cutoff (0.001 MeV) was 8876 min; 866 min with 0.01 MeV cutoff; 134 min with 0.1 MeV cutoff; and 43 min with 1 MeV cutoff. Using an electron energy cutoff of 1 MeV reduces the time required to complete the MCNP5 run by 99.5%; therefore, this cutoff value was used in this study.

3.3. Radiation environment during operation of M6 linac

While it is important to evaluate the dose rates within the facility due to operation of the M6 linac under normal operating conditions, it is also important to understand the dose rates for other possible scenarios. As future research activities may require the M6 linac usage without the fan beam collimators, it is necessary to evaluate the dose rates within the building under such operating conditions. Further, an understanding of the maximum dose rates achievable

	E (MeV)	< 10°	10–20°	20–30°	30–40°	40–50°	50–60°	60–70°	70–80°	80–90°
Default	0.1	4.47E − 03	4.11E − 03	3.39E − 03	2.52E − 03	1.59E − 03	7.91E − 04	2.76E − 04	5.31E − 05	8.56E − 07
Cut-off	0.1	3.58E − 03	3.28E − 03	2.68E − 03	1.97E − 03	1.24E − 03	6.00E − 04	2.02E − 04	3.70E − 05	6.24E − 07
Default	0.16337	1.29E − 02	1.12E − 02	9.29E − 03	7.04E − 03	4.87E − 03	2.75E − 03	1.15E − 03	2.57E − 04	3.58E − 06
Cut-off	0.16337	1.14E − 02	9.99E − 03	8.17E − 03	6.22E − 03	4.17E − 03	2.34E − 03	9.64E − 04	2.13E − 04	2.95E − 06
Default	0.22673	2.00E − 02	1.60E − 02	1.24E − 02	9.29E − 03	6.25E − 03	3.65E − 03	1.56E − 03	3.48E − 04	4.70E − 06
Cut-off	0.22673	1.88E − 02	1.51E − 02	1.17E − 02	8.50E − 03	5.66E − 03	3.25E − 03	1.40E − 03	3.07E − 04	4.12E − 06
Default	0.2901	2.05E − 02	1.59E − 02	1.18E − 02	8.56E − 03	5.78E − 03	3.35E − 03	1.48E − 03	3.37E − 04	4.53E − 06
Cut-off	0.2901	2.00E − 02	1.51E − 02	1.13E − 02	8.07E − 03	5.35E − 03	3.10E − 03	1.35E − 03	3.05E − 04	4.07E − 06

Table 1. Electron energy cutoff study: photon flux results.

http://dx.doi.org/10.5772/intechopen.71317

due to the M6 linac operation helps to characterize safety features within the building as well as evaluate the effectiveness of the linac shielding. The normal operation mode of the M6 linac includes the use of tungsten collimator pieces to shape the emitted x-rays into a horizontal fan beam at a height of 1.2 m above the floor. Lead shielding was used within the linac assembly to minimize dose rates to the sides and rear.

Three M6 linac configurations were studied: (1) normal operation configuration with collimators and shielding; (2) collimators were removed, but shielding left intact; (3) both collimators and shielding were removed (the maximum dose rate scenario). For each linac configuration, the FMESH tally was used to determine the overall dose rate footprint while F5 tallies were used to determine the dose rates at specific building locations. Comparison of the specific dose rates under differing M6 configurations allowed for determination of the effectiveness of the collimator pieces and linac shielding in reducing dose rates throughout the building. The MCNP5 models do not incorporate the earthen berm to the north east of the facility. This allows for studying the shielding effectiveness of the concrete wall alone. In actuality, the earthen berm completely envelops the northern, northeastern and eastern walls of the facility.

The computed dose rates due to the M6 linac operation under the normal operation configuration, in both high and low energy mode are shown in **Figures 6** and **7**. It was found that dose rates within the accelerator facility are higher when the M6 linac is operated in the high energy mode than when it is operated in the low energy mode. This is due to higher energy photons being produced (the endpoint energy of 6 MeV as opposed to 3 MeV) as well as larger fluxes of lower energy photons. For example, the computed flux of 1 MeV photons

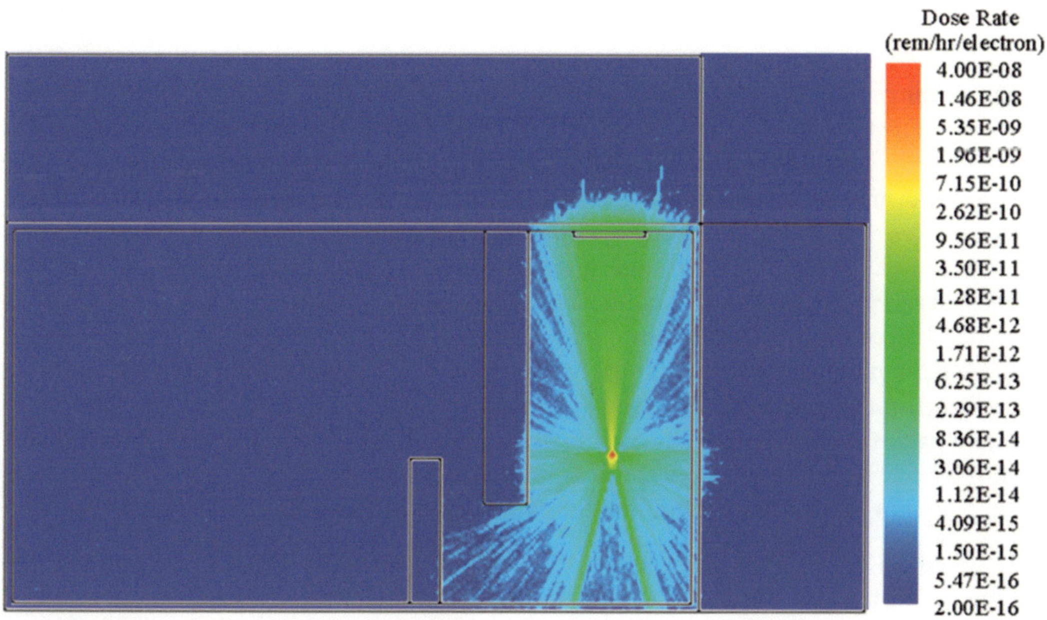

Figure 6. M6 linac in the normal operation configuration: dose rates for 6 MeV electron beam.

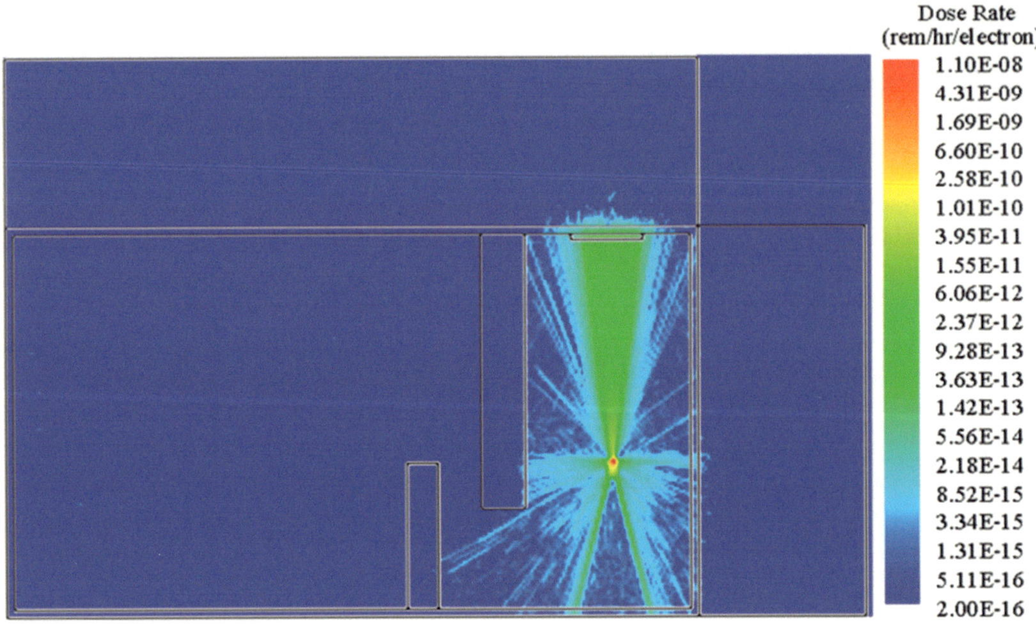

Figure 7. M6 linac in the normal operation configuration: dose rates for 3 MeV electron beam.

produced in high energy mode is an order of magnitude larger than when in low energy mode. The dose rates are largest directly in front of the linac, where the collimated beam is located. The fan shape is visible in both energy modes, with dose rates being higher in high energy mode. In both energy modes, the shielding maze minimizes the dose rates within the accelerator entry way.

Under certain conditions (i.e. production of photoneutrons using a low-threshold neutron converter or irradiation of large samples), the M6 linac may be used without the tungsten collimator pieces. As these collimator pieces attenuate the majority of emitted photons in all but the specific beam shape, the removal of these pieces leads to an increase in the photon fluxes and dose rates expected not only in the northern half of the facility, but throughout the building. The emitted photons will no longer take the shape of a fan beam, but rather a cone with dimensions according to the collimator cavity. The expected dose rates due to the M6 linac operation without collimators are shown in **Figures 8** and **9**. Similar to the configuration 1, the dose rates in the accelerator facility when the M6 linac is operated without collimators are higher when it is operated in the high energy mode. It was found that the dose rates within the northern half of the accelerator bay are greatly increased when the collimators are removed. In addition, it was determined that the dose rates in the entry way (0.259 ± 0.0020 rem/h) were larger than they were in the configuration 1 (0.0004 ± 0.00002 rem/h).

Determination of the dose rates within the accelerator facility for operation of the M6 without shielding and collimator materials constitutes the "worst case scenario," or maximum possible dose rates achievable (see **Figures 10** and **11**). It is important to evaluate these dose rates in order to help validate safety measures for the facility. Removing the

http://dx.doi.org/10.5772/intechopen.71317

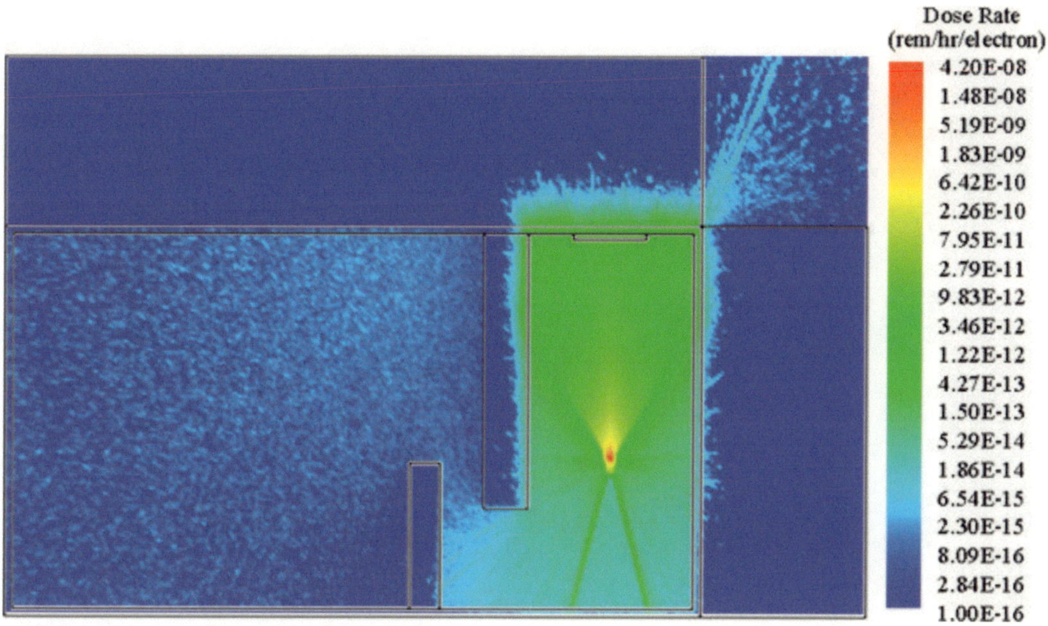

Figure 8. M6 linac in the 2nd configuration: dose rates for 6 MeV electron beam.

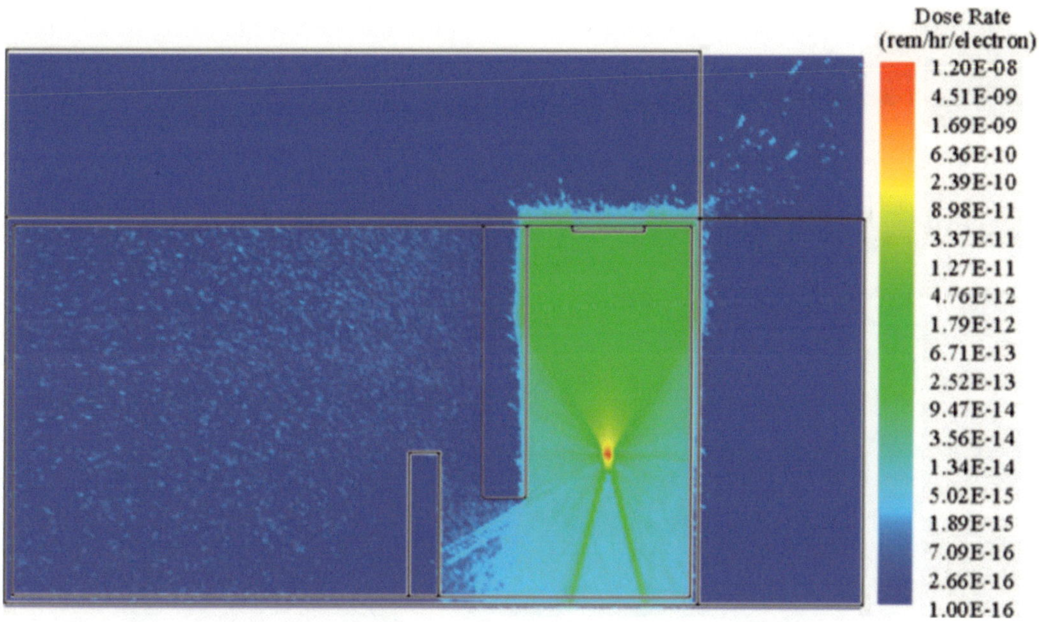

Figure 9. M6 linac in the 2nd configuration: dose rates for 3 MeV electron beam.

shielding and collimators results in an increase in the dose rates throughout the facility. When compared to the results from configurations 1 and 2, it was found that the dose rates in the accelerator bay to the rear and sides of the linac increased by an order of magnitude.

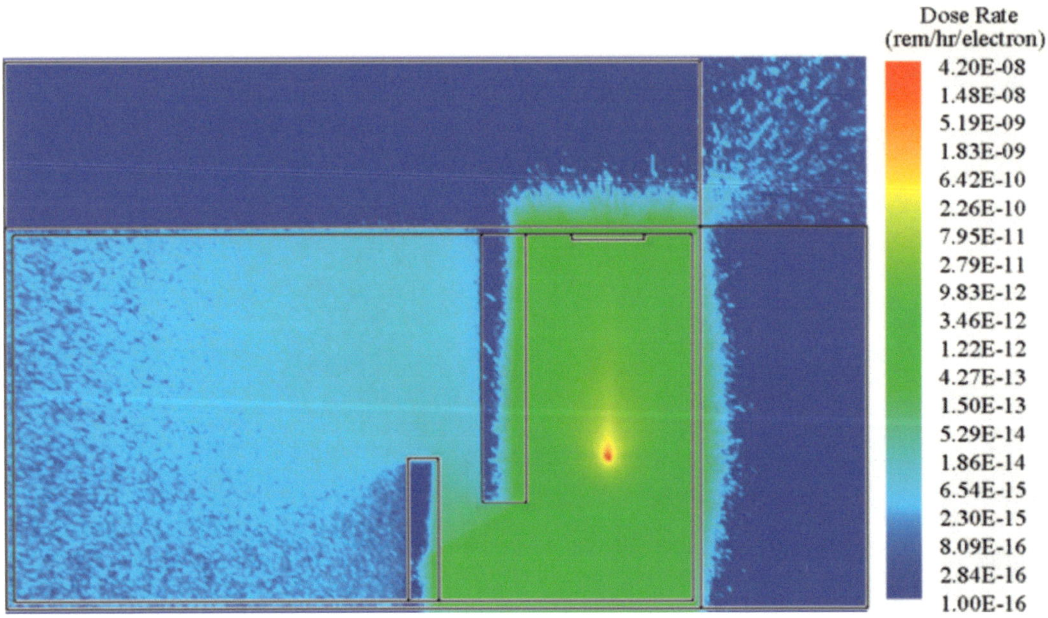

Figure 10. M6 linac without shielding and collimators: dose rates for 6 MeV electron beam.

This is due to the removal of the lead shielding in the rear of the M6 linac. In addition, the dose rates increased within the shielding maze (8.9 ± 0.05 rem/h) as well as the entry way (1.4 ± 0.005 rem/h). A summary of results detailing the product of the F5 dose rate

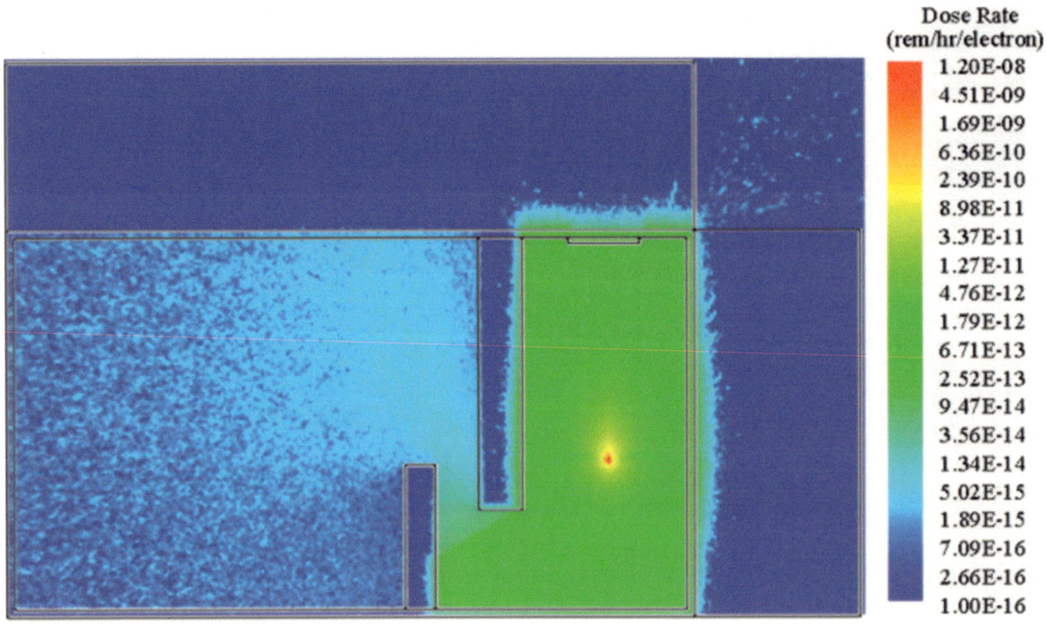

Figure 11. M6 linac without shielding and collimators: dose rates for 3 MeV electron beam.

http://dx.doi.org/10.5772/intechopen.71317

tallies with the M6 electron current for all three configurations of the M6 linac operation is shown in **Table 2**. These results quantify the trends from the dose rate maps (shown in **Figures 6–11**) at specific building locations.

At all tally locations, the lowest dose rates occur under the normal M6 linac operation mode while the maximum dose rates occur when the collimators and shielding have been removed. When M6 collimators and shielding are present, the dose rates in the corners of the northern bay are reduced by a factor of 265 while the dose rates in the southern corners are reduced by a factor of 105. The entry way dose rates are reduced by a factor of 350 while the dose rate in the center of the shielding maze is reduced by a factor of 180. In the absence of the berm outside the northeast corner of the building, the dose rate was found to be under 3 rem/h. When the berm is present, the dose rates outside fall below the 10 Code of Federal Regulations (CFR) 20 limit, for the dose rate in an unrestricted area (2 m rem/h). At 2 m north of the linac, F5 tallies were used to determine the vertical dose rate profile for all three operating configurations (shown in **Figure 12**).

The results reflected large dose rates consistent with a fan beam shape at 1.2 m above the floor during normal operation. The use of collimators was shown to reduce the dose rates at all tally locations except at fan beam level. When collimators were removed, the fan beam expands to a cone shape and the dose rate increases. It was found that at dose rates near the ceiling (above 250 cm) were slightly lower (3.5×10^3 rem/h) due to the tally locations being outside the radiation cone beam. During operation of the M6 linac, the dose rate is continuously measured and monitored by an internal ion chamber calibrated to a distance of 1 m north of the linac. When the M6 linac was operated in the 6 MeV mode, the dose rate was measured to be 2.44×10^4 rem/h. The computational dose rate was determined by multiplying the normalized F5 tally result by the electron current and found to be 2.76×10^4 rem/h. The model and experimental measurement were found to be in agreement, with MCNP5 providing a conservative estimate for photon dose that is 1.13 time the measured value.

Facility location	**Dose rate (rem/h)**		
	Normal operation	**Without collimators**	**Maximum doses**
Entry room	0.0004 ± 0.00002	0.259 ± 0.0020	1.4 ± 0.005
Shielding maze	0.002 ± 0.0001	0.317 ± 0.0046	8.9 ± 0.05
Northern corners	3.29 ± 0.16	556 ± 1.3	583 ± 1.2
Southern corners	0.037 ± 0.0008	6.3 ± 0.07	206 ± 0.5
At 1 m	27,571 ± 69	42,081 ± 93	41,986 ± 88
Sample table	741 ± 1.93	1359 ± 2.7	1396 ± 2.8
Outside (no berm)	n/a	n/a	2.9 ± 0.08

Table 2. Accelerator facility dose rates caused by the M6 linac operation.

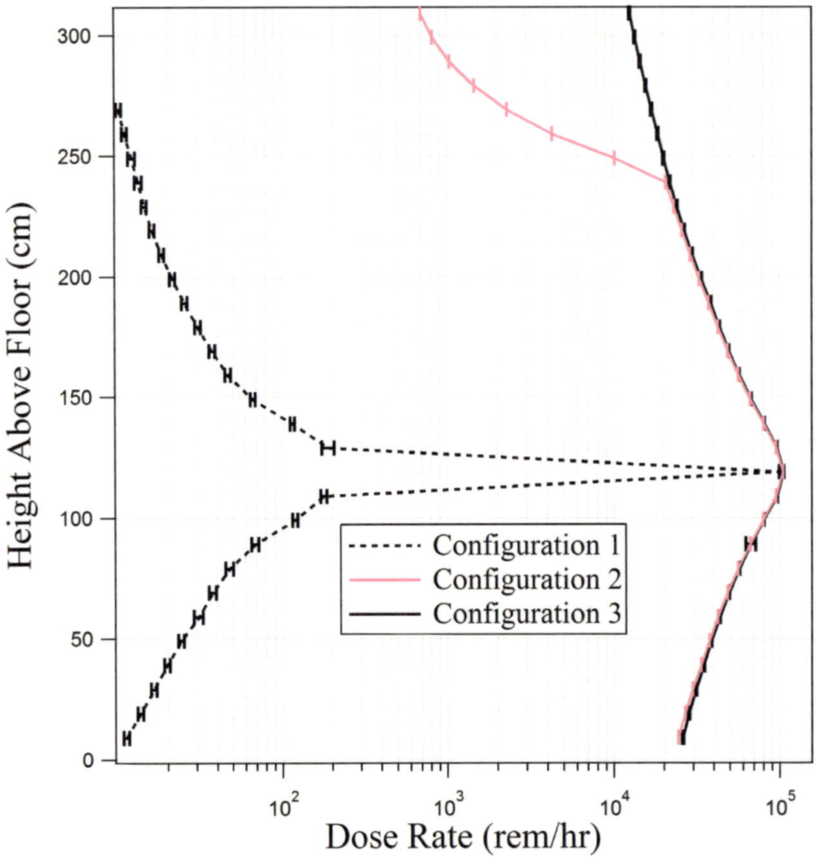

Figure 12. Vertical profile of dose rate during the M6 linac operation.

3.4. Radiation environment during operation of K15 linac

The K15 linac is not typically operated with a fan beam collimator, but rather with a 30° cone beam collimator. Other collimators may be used, but for the purpose of this study, the 30° cone case was considered. Two scenarios were modeled for the K15 linac operation: (1) normal operation with the cone collimator and shielding; (2) operation without shielding or collimators (the maximum dose rate scenario).

For each linac operating configuration, the TMESH/RMESH tally was used to determine the overall dose rate footprint in the building. Due to the energies of the photons generated in the linac target in the high energy mode being greater than the neutron binding energies of some materials in the linac shielding as well as the facility room structures, the photoneutron flux as well as its contribution to the dose rate must be considered. For photon energies higher than 10 MeV, photoneutron generation was expected. For example, photoneutron thresholds for isotopes in some materials are the following: 10.56 MeV for ^{14}N, 13.06 MeV for ^{27}Al, 10.23 MeV for ^{55}Mn, 11.20 MeV for ^{56}Fe, 12.22 MeV for ^{58}Ni, 10.85 MeV for ^{63}Cu, and 11.86 MeV for ^{64}Zn.

Under normal operation of the K15 linac, the photon dose rates were computed and shown in **Figures 13** and **14** for the low and high energy modes, respectively. The trends in the computed dose rates due to the K15 operation were found to be similar to those due to the M6 linac operation. The dose rates in the accelerator bay were highest, while the shielding maze helped to minimize dose rates in the entry way. The earthen berm minimized the photon dose rate outside of the facility to effectively nothing. Dose rates to the sides and rear of the K15 linac were higher when operated in high energy mode as compared to low energy mode. The vertical profile of the photon dose rate at 1 m north of the linac was measured using F5 tallies and is shown in **Figure 15**. The error associated with each value is less than the 5% recommended by MCNP for F5 tallies.

The computed normalized results show that the photon dose rate is largest down the axis of the collimated photon beam (9.5×10^{-8} rem/h/electron), at a height of 1.11 m above the floor. The dose rates near the floor (9.0×10^{-12} rem/h/electron) and the ceiling (1.0×10^{-8} rem/h/electron) of the building were determined to be approximately four orders of magnitude lower than the dose rate in the beam axis. The largest dose rates were found to occur between heights of 75 and 125 cm, corresponding to the height of the conical collimated photon beam. Dose rates quickly decrease outside of the photon beam.

The neutron contribution to the dose rate during the normal K15 linac operation in the high energy mode is shown in **Figure 16**. No photoneutrons were produced during operation of the K15 linac in the low energy mode due to the endpoint energy of the bremsstrahlung being below the (γ,n) reaction thresholds of the materials in the MCNPX model.

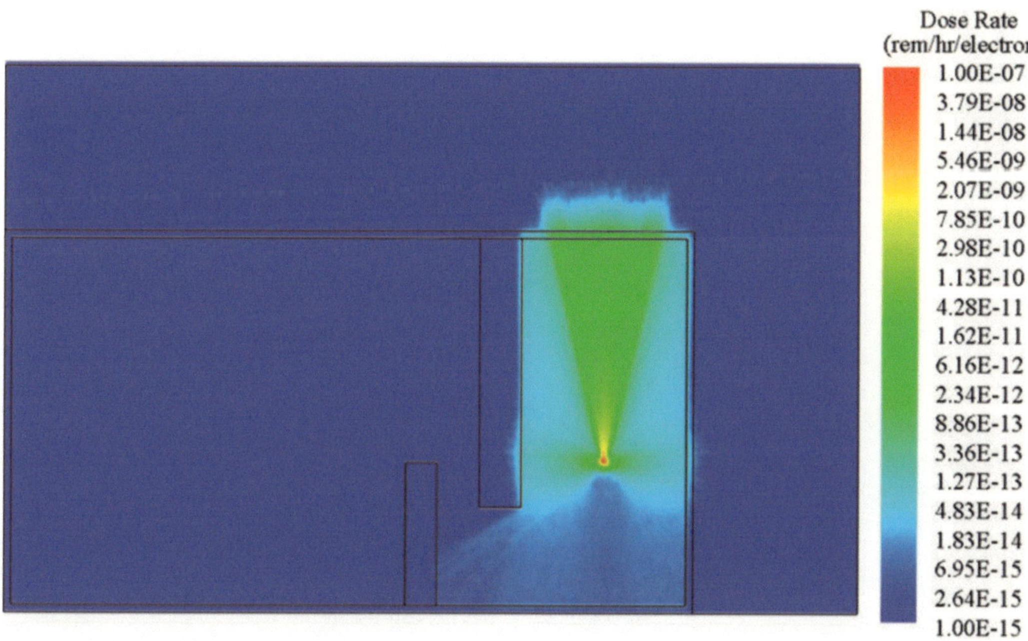

Figure 13. K15 linac in configuration 1: photon dose rates for 9 MeV electron beam.

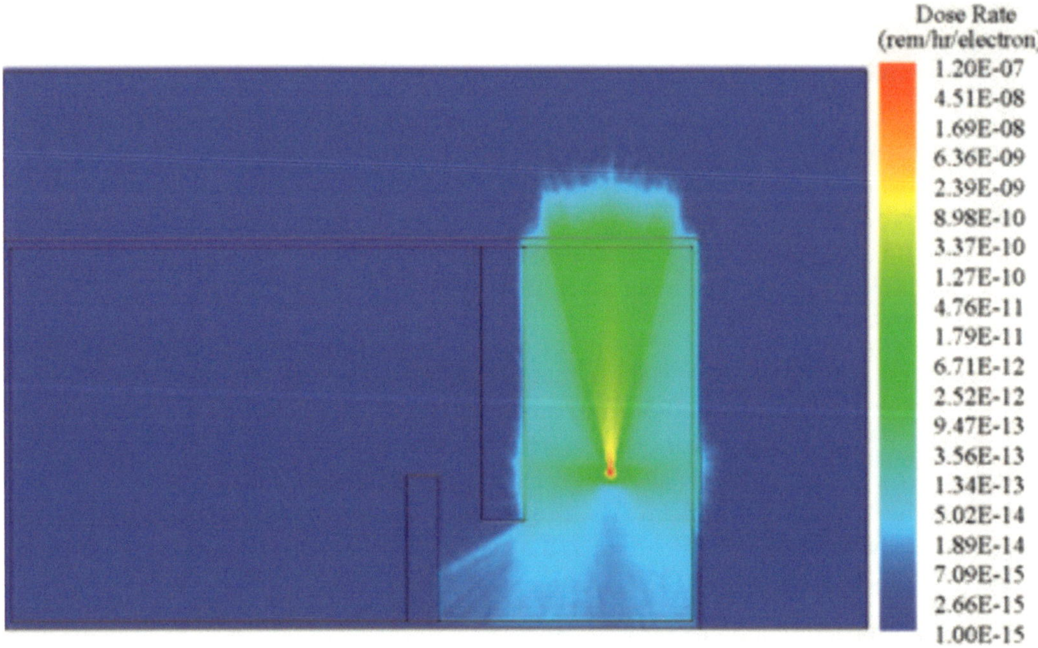

Figure 14. K15 linac in configuration 1: photon dose rates for 15 MeV electron beam.

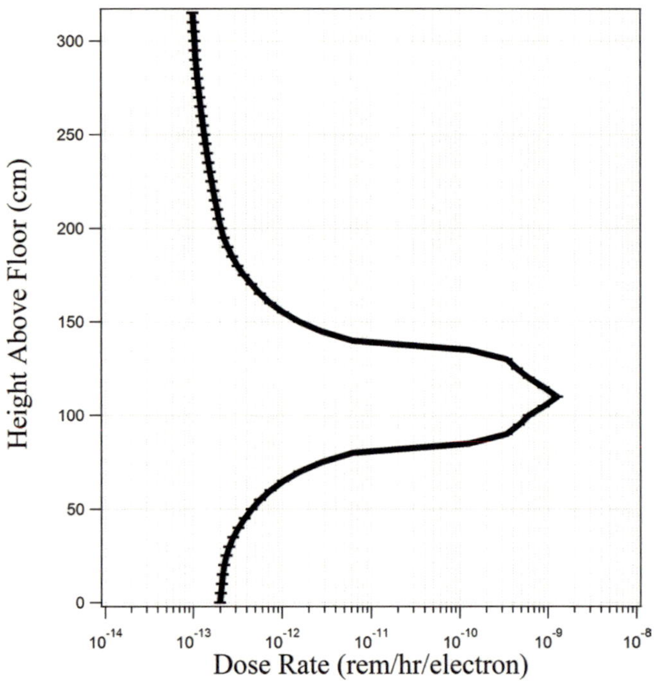

Figure 15. Vertical profile of photon dose rate in normal operation of K15 linac in high energy mode.

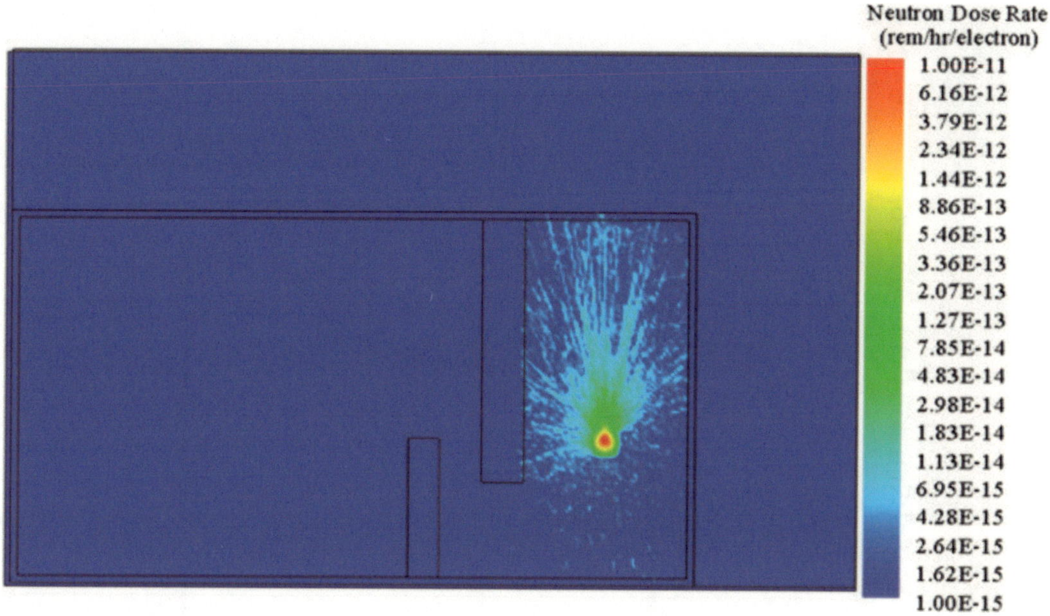

Figure 16. K15 linac in configuration 1: neutron dose rates for 15 MeV electron beam.

The results show that the neutron flux was primarily contained within the accelerator bay. Neutron contribution to dose rate was highest north of the linac primarily due to the lack of low-Z shielding behind the target head. The back end of the K15 linac contained several inches of polyethylene shielding which reduced the neutron dose rate in the southern end of the bay. The maximum dose rate due to neutron flux was determined to be several orders of magnitude lower than the photon contribution. While the profile shape of the photon dose rate corresponded to the shape of the conic collimator, the neutron dose rate does not possess the same shape. This is because neutrons were produced in the high-Z collimator materials rather than being shaped by it. The neutron spectrum at a distance of 1 m behind of the linac target was computed using an F5 tally (see **Figure 17**). The largest flux of neutrons was determined to be in the 0.1–1 MeV range with the second largest flux for neutrons just above the thermal range (10^{-8}–10^{-7} MeV). The total neutron flux at the F5 tally location was found to be 4.8×10^4 neutrons/cm^2/s.

The geometry in the MCNPX model was modified to simulate the K15 linac operation without shielding and collimators in order to determine the maximum dose rate due to operation of the linac in the high energy mode. Maximum photon dose rate results are shown in **Figure 18**. The results show that without collimators and shielding, the dose rates due to photons increase throughout the building. Comparing the RMESH tally data between the two linac operating configurations reveals that the K15 collimator and shielding materials help to reduce photon dose rates by factors of 238, 33 and 7.5 times for locations at 1 m north of the linac, in the center of the shielding maze and in the center of the entryway, respectively.

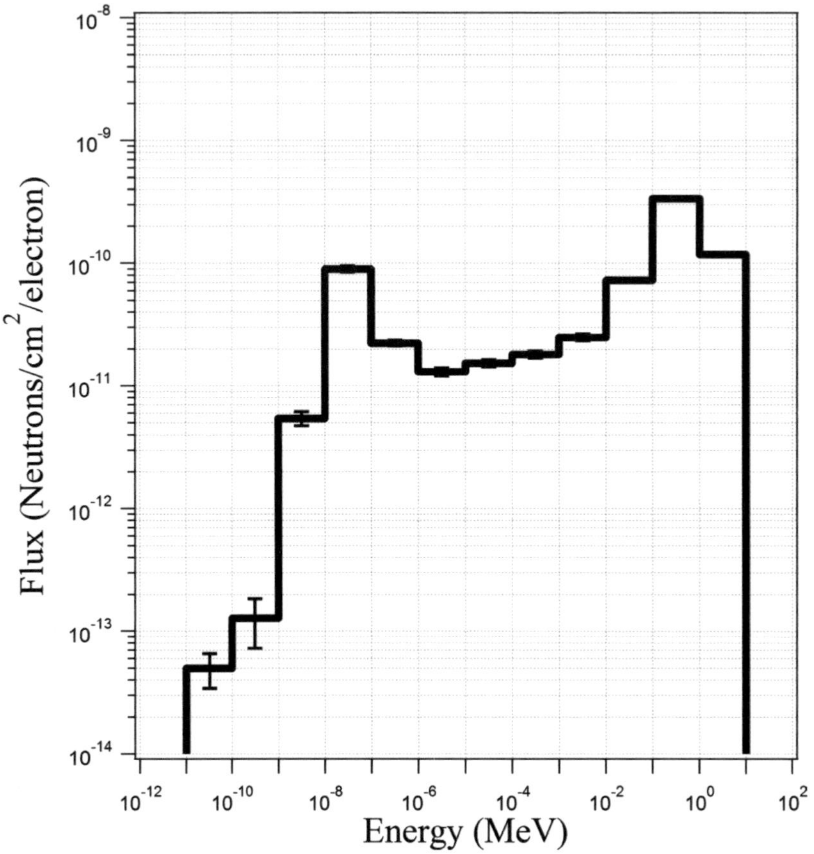

Figure 17. Photoneutron spectrum at 1 m from the K15 linac target.

The maximum neutron dose rate footprint is similar to that of the normal neutron dose rate map with the exception that the dose rate has increased in the southern half of the accelerator bay. This is due to the fact that the polyethylene neutron shielding was removed in the rear of the linac.

The photon dose rate from the K15 linac was experimentally measured at Varian Medical Systems by an internal ion chamber (calibrated at 1 m north of the linac) and found to be 11,700 rem/min for high energy mode and 3500 rem/min for low energy mode. MCNPX F5 tally results at the same locations yielded values of 1.40×10^{-9} and 3.10×10^{-10} rem/h/starting electron for high and low energy respectively. Multiplying these tally values by the respective DC averaged electron currents and converting to the appropriate time scale provided values for high and low energy dose rate as 15,744 and 3687 rem/min, respectively. Thus, the MCNPX model provided conservative estimates of photon dose rate by scale factors of 1.35 for the high energy mode and 1.05 for the low energy mode.

It is important for safety purposes to evaluate differences in the radiation production between the two linacs that can be used at the same facility. At 1 m north of the linac, the M6 linac generates a photon dose rate of over 400 rem/min. At the same location, the photons from the K15 linac generate dose rates 28 times larger (high energy mode) and 8 times larger (low energy

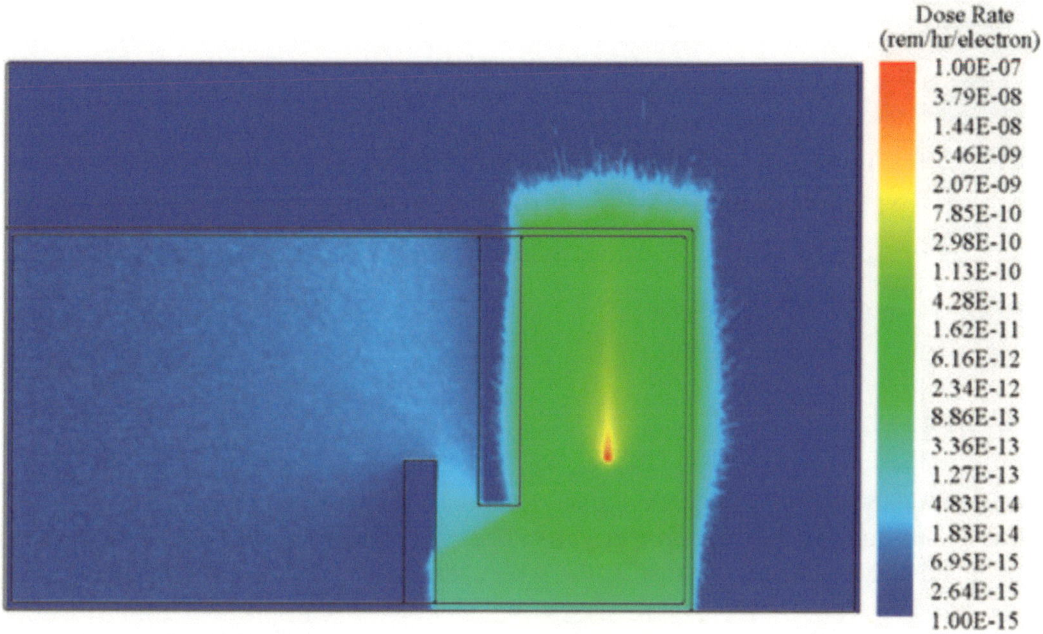

Figure 18. K15 linac without shielding and collimators: photon dose rate for 15 MeV electron beam.

mode) than the M6 linac does. When compared to the M6 results, the K15 maximum dose rates an order of magnitude larger. At 1 cm behind the respective linac targets, the maximum photon flux occurs within a 10° conic angle. At this angle, the total photon flux (normalized computational result multiplied by the electron current) of the K15 linac in the high energy mode is over four times as large as that of the M6 linac (**Table 3**).

Linac model (Electron energy)	K15 (15 MeV)	K15 (9 MeV)	M6 (6 MeV)
	Photon Flux (photons/cm²/s)		
Angle, degrees			
10	4.27E + 14	1.47E + 14	9.53E + 13
20	2.81E + 14	1.07E + 14	7.00E + 13
30	1.89E + 14	7.66E + 13	4.93E + 13
40	1.27E + 14	5.42E + 13	3.40E + 13
50	8.03E + 13	3.61E + 13	2.19E + 13
60	4.38E + 13	2.09E + 13	1.24E + 13
70	1.78E + 13	8.97E + 12	5.41E + 12
80	4.34E + 12	2.60E + 12	1.29E + 12
90	9.75E + 10	6.91E + 10	2.04E + 10

Linac model (Electron energy)	K15 (15 MeV)	K15 (9 MeV)	M6 (6 MeV)
	Does Rate (rem/m/min)		
Measured dose rate	11,700	3500	406.67
MCNP computed dose rate	15,744	3687	460
Scale factor (MCNP dose/measured dose)	1.35	1.05	1.13

Table 3. Characteristics of M6 and K15 linacs.

4. Conclusion

Radiation safety aspects of operation of electron linacs equipped with bremsstrahlung converters were studied. The operation of Varian linacs M6 (3 and 6 MeV electron beams) and K15 (9 and 15 MeV electron beams) at the accelerator facility of University of Nevada, Las Vegas was investigated. High-energy photon and photoneutron production during the linac operation were analyzed using Monte Carlo computational models. The bremsstrahlung spectral and angular distributions were computationally determined for the M6 linac and the K15 linac. Biological equivalent dose rates due to accelerator operation were evaluated using the flux-to-dose conversion factors. Dose rates were computed for the accelerator facility for the linac use under different operating conditions. The results showed that the use of collimators and the linac's internal shielding significantly reduced the dose rates throughout the facility. Measurements shown that computational models allow conservative dose rate estimations.

Photoneutron dose rates within the facility were computed for the K15 linac in the high energy mode. While the largest neutron contribution to dose rate was four orders of magnitude lower than the photon contribution, it was still large enough to warrant consideration when designing the bunker shielding.

Author details

Matthew Hodges* and Alexander Barzilov

*Address all correspondence to: Hodgesm@unlv.nevada.edu

University of Nevada, Las Vegas, United States

References

[1] Wangler T. RF Linear Accelerators. 2nd ed. Weinheim: Wiley-VCH Verlag; 2008. 466 p. DOI: 10.1002/9783527623426

[2] Broemme J et al. Adjuvant therapy after resection of brain metastases: Frameless image-guided LINAC-based radiosurgery and stereotactic hypofractionated radiotherapy. Strahlentherapie und Onkologie. 2013;**189**(9):765-770. DOI: 10.1007/s00066-013-0409-z

[3] Fong BM et al. Hearing preservation after LINAC radiosurgery and LINAC radiotherapy for vestibular Schwannoma. Journal of Clinical Neuroscience. 2012;**19**:1065-1070. DOI: 10.1016/j.jocn.2012.01.015

[4] Brice S. Proton improvement plan II: An 800 MeV superconducting linac to support megawatt proton beams at Fermilab. Nuclear and Particle Physics Proceedings. 2016;**273-275**:238-243. DOI: 10.1016/j.nuclphysbps.2015.09.032

[5] Brunner O et al. Assessment of the basic parameters of the CERN superconducting proton linac. Physical Review Special Topics - Accelerators and Beams. 2009;**12**:070402-1-070402-24. DOI: 10.1103/PhysRevSTAB.12.070402

[6] Kim HJ et al. Superconducting linac for the rare isotope science project. Journal of the Korean Physical Society. 2015;**66**(3):413-418. DOI: 10.3938/jkps.66.413

[7] Jones J et al. Detection of shielded nuclear material in a cargo container. Nuclear Instruments and Methods in Physics Research A. 2006;**562**:1085-1088. DOI: 10.1016/j.nima.2006.02.101

[8] Barzilov A, Womble PC, Vourvopoulos G. NELIS – A neutron inspection system for detection of illicit drugs. AIP Conference Proceedings. 2003;**680**:939-942. DOI: 10.1063/1.1619863

[9] Chen G, Bennett G, Perticone D. Dual-energy X-ray radiography for automatic high-Z material detection. Nuclear Instruments and Methods in Physics Research B. 2007;**261**:356-359. DOI: 10.1016/j.nimb.2007.04.036

[10] Hartman J et al. 3D imaging using combined neutron-photon fan-beam tomography: A Monte Carlo study. Applied Radiation and Isotopes. 2016;**111**:119-116. DOI: 10.1016/j.apradiso.2016.02.018

[11] Kosako K et al. Angular distributions of Photoneutrons from copper and tungsten targets bombarded by 18, 28, and 38 MeV electrons. Journal of Nuclear Science and Technology. 2011;**48**(2):227-236. DOI: 10.1080/18811248.2011.9711696

[12] U.S. NRC. Units of Radiation Dose [Internet]. 2017. Available from: http://www.nrc.gov/reading-rm/doc-collections/cfr/part020/part020-1004.html [Accessed: Aug 25, 2017]

[13] ANSI Standard ANSI/ANS 6.1.1-1977. Neutron and Gamma-ray Flux-to-Dose-Rate Factors. La Grange, Illinois: American Nuclear Society; 1977

[14] Kirkby C et al. RBE of KV CBCT radiation determined by Monte Carlo DNA damage simulations. Physics in Medicine and Biology. 2013;**58**(16):5693-5704. DOI: 10.1088/0031-9155/58/16/5693

[15] Heerikhuisen J, Florinski V, Zank GP. Interaction between the solar wind and interstellar gas: a comparison between Monte Carlo and fluid approaches. Journal of Geophysics Research: Atmosphere. 2006;**111**(A6):1-8. DOI: 10.1029/2006JA011604

[16] Creal D. A survey of sequential Monte Carlo methods for economics and finance. Economic Review. 2012;**31**(3):245-296. DOI: 10.1080/07474938.2011.607333

[17] Barzilov A, Novikov I, Cooper B. Computational study of pulsed neutron induced activation analysis of cargo. Journal of Radioanalytical and Nuclear Chemistry. 2009;**282**:177-181. DOI: 10.1007/s10967-009-0298-x

[18] Barzilov A, Womble PC. Study of Doppler broadening of gamma-ray spectra in 14-MeV neutron activation analysis. Journal of Radioanalytical and Nuclear Chemistry. 2014;**301**:811-819. DOI: 10.1007/s10967-014-3189-8

[19] Brown F, Kiedrowski B, Bull J. MCNP5 1.60 Release Notes. Los Alamos National Laboratory. Los Alamos, New Mexico: LA-UR-10-06235; 2010

[20] Pelowitz D. MCNPX 2.7.0 Extensions. Los Alamos National Laboratory. Los Alamos, New Mexico. LA-UR-11-02295; 2011

[21] McConn RJ et al. Compendium of Material Composition Data for Radiation Transport Modeling. Richland, Washington: Pacific Northwest National Laboratory. PNNL-15780; 2011

[22] Verbeke J et al. Simulation of Neutron and Gamma Ray Emission from Fission and Photofission. Livermore, California: Lawrence Livermore National Laboratory. UCRL-AR-228518; 2014

[23] Hughes HG. Treating Electron Transport in MCNP. Los Alamos National Laboratory. Los Alamos, New Mexico: LA-UR-96-4583; 1997

Section 4

Accelerator Applications

X-Ray Diffraction Detects D-Periodic Location of Native Collagen Crosslinks *In Situ* and Those Resulting from Non-Enzymatic Glycation

Rama Sashank Madhurapantula and
Joseph P.R.O. Orgel

Additional information is available at the end of the chapter

http://dx.doi.org/10.5772/intechopen.71022

Abstract

Synchrotron based X-ray diffraction experiments can be highly effective in the study of mammalian connective tissues and related disease. It has been employed here to observe changes in the structure of Extra-Cellular Matrix (ECM), induced in an *ex vivo* tissue based model of the disease process underlying diabetes. Pathological changes to the structure and organization of the fibrillar collagens within the ECM, such as the formation of non-enzymatic crosslinks in diabetes and normal aging, have been shown to play an important role in the progression of such maladies. However, without direct, quantified and specific knowledge of where in the molecular packing these changes occur, development of therapeutic interventions has been impeded. *In vivo*, the result of non-enzymatic glycosylation i.e. glycation, is the formation of sugar-mediated crosslinks, aka advanced glycation end-products (AGEs), within the native D-periodic structure of type I collagen. The locations for the formation of these crosslinks have, until now, been inferred from indirect or comparatively low resolution data under conditions likely to induce experimental artifacts. We present here X-ray diffraction derived data, collected from whole hydrated and intact isomorphously derivatized tendons, that indicate the location of both native (existing) and AGE crosslinks *in situ* of D-periodic fibrillar collagen.

Keywords: type I collagen, diabetes, crosslinking, advanced glycation end-products, X-ray diffraction

1. Introduction

The extracellular matrix (ECM) is a complex network of biomolecules that provides structural and functional support to cells. Both within and with the ECM, cells proliferate and form three-dimensional tissues in multicellular organisms. As such, the ECM is important to homeostasis, function and survival in higher vertebrates. Varying composition of the ECM per tissue, supports widely different biological functions. For instance, blood vessels, organ systems, bones, tendons, skin and so forth although significantly different in material properties and cellular-tissue functions, contain many common ECM elements. The collagens (there are nearly 30 different types) account for a major part of nearly all vertebrate ECMs. Type I collagen is the most predominant collagen found in animals (and also found in some insects and plants). It is the major constituent of tendons, ligaments, skin, bone, teeth, body organ frameworks and so forth. While type I collagen is surpassed in some specific connective tissues (such as in cartilage), major structural features of type I collagen appear to be largely common to the other fibril-forming types [1]. Therefore, the research presented here, conducted on type I collagen rich tendons, can be used to extend our structural understanding of other fibrillar collagens and tissues with fairly reasonable extrapolation. This facility is enhanced by the fact that the structure of type I collagen is one of the best-defined structures in the collagen family. The *in situ* packing structure of type I collagen from rat-tail tendons has been determined with an associated and publicly accessible molecular model resulting from this work [2] (**Figure 1**). With this structural model and the data resulting from this study, the effects of molecular changes in collagen

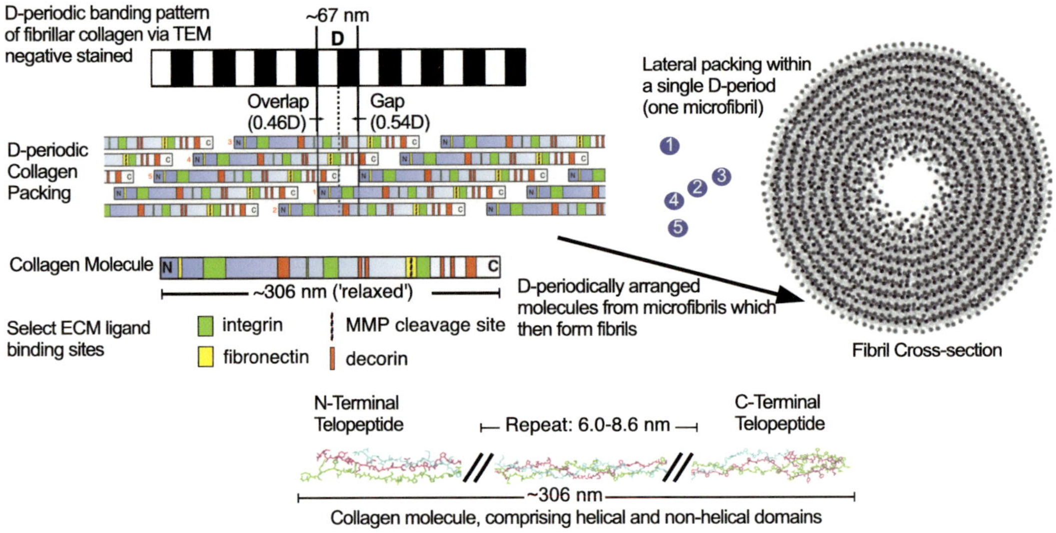

Figure 1. Structural hierarchy of type I collagen. The principal components of collagen fibrils are the triple helices (~306 nm) that are primarily composed of peptides with a residue sequence of G-X-Y (where X and Y are commonly Proline and Hydroxyproline). The triple helical regions of the monomers are flanked on the N and C termini by non-helical telopeptides. Neighboring collagen molecules are staggered from one another to form the collagen microfibril. This is the three-dimensional structure formed by the 67 nm D-periodic repeat comprising an overlap region (where 5 monomer molecular segments align) and the gap region (4 monomer molecular segments align). Several important ligand binding sites are indicated (colored boxes within collagen molecules) on the monomers which adopt a network of potentially related sites in the microfibrillar packing of collagen.

structure and organization brought about by disease may be understood and thus treatments formulated to address them.

1.1. Type I collagen – molecular packing and stabilization

Type I collagen is the major component of rat-tail tendons, the sample tissue of this study. These tendons samples are not single crystals; their supramolecular arrays are very large in cellular terms and are insoluble. Therefore, normal (single-crystal) crystallographic methods cannot be applied without adaptation. Nor is it necessary to form artificial crystals of collagen as the collagen molecules *in situ* of their tissues are *crystalline*. They are arranged in a quasi-hexagonal packing scheme radially and D-periodic staggered packing scheme longitudinally that form supramolecular structures [1], the microfibril and fibril. The fundamental biological units of this ECM protein in animals are the collagen fibrils, which also are the crystalline units that diffract X-rays. These fibrils align nearly parallel to each other within tendons (**Figure 1**). As alluded to above, the ~300 nm long collagen molecules align axially to form the 67 nm staggered repeat (with 234 amino acids per molecular segment within this 67 nm repeat) as observed by electron microscopy and X-ray diffraction. This 67 nm repeat is referred to as the D-period where fractions of D provide an axial location originating from the N-terminal end of the D-period or referring to how much of the D-period is gap or overlap. Within these 67 nm repeats, are the overlap region which is 0.46D and the gap region which is 0.54D.

The axial (or longitudinal) arrangement of collagen molecules within a fibril is significantly more crystalline than the lateral (or radial) packing discussed above, giving a series of well-ordered Bragg X-ray reflections along the meridian of the X-ray fiber diffraction pattern [3]. Putting the information from *both* the lateral and axial X-ray diffraction, Orgel et al. [2] reported a full 3D structure and accompanying model that is well-accepted and regularly forms the basis of structural-function interpretation from the molecular scale upwards.

The hierarchical packing structure explained above is maintained *in vivo* by means of post-translational modifications to the collagen fibril and through the attachment of stabilizing ligands to both the fibril and fibril-bundle(s) [4, 5]. In this present study, we look at the role of chemical crosslinks on the axial packing structure of collagen and the specific placement within the D-period of these cross-links using X-ray diffraction methodology.

1.2. X-ray fiber diffraction

The principals of X-ray crystallography have been previously applied to study the structure of types I and II collagen and how they are adapted to their functional role [1, 2, 6, 7].

As is well known, X-rays are a form of electromagnetic radiation. Much as when visible light is incident on an object, the object scatters these rays. For visible light, the scattered rays can then be refocused using a lens and these focused rays form an image. For scattered X-rays however although theoretically possible, image focus, is in practice not yet achievable for the study of biological samples on the molecular scale. However, as a matter of almost common course at this time, the diffracted (scattered) X-rays can be recorded using X-ray detectors such as charge coupled devices (CCDs).

Rat-tail tendons diffract focused X-rays to give rise to a series of well-ordered Bragg reflections in alignment with the long axis of the tendon, which can be recorded on CCDs. As mentioned earlier, these reflections are a result of the crystalline D-periodic packing of collagen along the axis [7, 8]. The amplitudes (the square root of recorded intensity) of the Bragg reflections extracted from these images can be used with appropriate scaling, along with the missing 'phase' information to calculate electron density maps, along one crystallographic unit cell [9, 10]. One such method, is to use the known phase information of another known structure; one that is isomorphous to the structure under investigation is particularly advantageous. This was the method employed for the analyses presented in this study.

Native phases from the previously determined structure of type I collagen were combined with the amplitude information obtained from XRD patterns of isomorphous derivatives of rat-tail tendons created by incubating them in sugar solutions to create non-enzymatically formed chemical crosslinks (analogous to those formed in diabetes and normal aging) and further derivatizing them with microscopy dyes specific to these crosslinks. Appropriate scaling of native and isomorphous derivative data can be used to calculate difference Fourier and Patterson maps that may be used to locate those differences induced in the experimental tendons (glycated tendons) and those that are native (representative of normal, non-diseased state). This is based on the same method that was used to determine the structure of type I collagen [1, 2, 7]. The phase information from this structure may then be used in creating difference Fourier maps that indicate the locations of chemical cross-links to a level of accuracy that is similar to the resolution of the data used to obtain the maps.

1.3. Fourier and Patterson maps

The diffraction pattern of an object is equivalent to its own Fourier transform. Hence a reverse Fourier transform calculated using the amplitudes of the Bragg reflections obtained from the XRD patterns and phases would result in a continuous electron density profile along one D-period (the axial crystallographic unit cell) of type I collagen. A "difference Fourier" can be calculated by using the subtracted difference between derivative (for instance glycated tendon) and native (unmodified tendon) diffraction amplitudes and the phases for the native structure. This difference Fourier provides a map of electron densities added to the native structure and thus indicate regions in the D-period where chemical modifications have occurred.

In addition, calculating the Patterson and difference Patterson function from glycated, dyed and unmodified rat-tail tendons gives data on the distribution and the magnitude of electron dense regions along the D-period of collagen. The use of the Patterson function provides an important cross-verification of the accuracy of the interpretation based on the difference Fourier map/s. The key supporting value of the Patterson function, is that it does not rely on phase information and is a 'pure' observation of key data. This also makes it much more complex in terms of the structure of the molecule, than the electron density maps reconstructed by Fourier maps. A molecule with N atoms will give rise to N(N−1)+1 peaks in a Patterson map. However, with the proper amino acid sequence information as we have for type I collagen and some knowledge of its structure (we have substantial detailed knowledge of its native structure), the Patterson map becomes a powerful instrument of model validation for fibrillar collagen structures based on X-ray diffraction data [6, 11].

A method of cross-verification between Difference Fourier and Difference Patterson maps has been established previously, by using the observed locations of changes in electron densities from the difference Fourier to synthesize a "model" Patterson map of the difference between derivative and native (i.e. where the crosslinks occur and have been labeled). This is then compared to the observed difference Patterson derived from the amplitudes of the derivative and native diffraction patterns. Since the Patterson map is completely unaffected by assumptions about phase information, a difference Patterson of derivative and native data should represent the standard to which a model Patterson should conform. The model Patterson should be similar to that of the (model independent) difference Patterson if the Difference Fourier and resulting interpretive model are correct (the interpretive model from the Fourier map being used to generate the model Patterson function). This approach was applied to the study of glycated tendons presented here.

1.4. Enzymatic crosslinks in type I collagen

Enzymatic crosslinking is the formation of stabilizing covalent linkages between collagen monomers in the molecular packing of collagen [12]. The mechanism of crosslinking is effectively one of the final steps in the building of fibrils, thereby functionalizing collagen [13]. This happens through a specific set of enzymatic reactions at determined locations. A post-translational enzymatic oxidative deamination of lysine or hydroxylysine residues by the enzyme lysyl oxidase results in the formation of intra and intermolecular crosslinks with other lysine or hydroxylysine residues (**Figure 2**). The activity of this enzyme is highly regulated by steric requirements and by the amino acid sequences upstream and downstream of the target [14]. The quantification of these crosslinks has been established using chromatography [15].

The locations of these enzymatic crosslinks constrain the possible conformation of the C-telopeptides of the two $\alpha 1$ chains to form sharp hairpin turns at around residues 13 and 14 of the 25-residue telopeptides [16]. The formation of intermolecular crosslinks at specific sites is a stabilizing mechanism for the supramolecular packing of collagen molecules. Impairments of crosslinks, either destabilization or hyper stabilization can lead to pathological conditions [17–19].

1.5. Glycation

Glycation, also known as Maillard reaction, is a non-enzymatic reaction between reducing sugars and amino groups in proteins, lipids and nucleic acids. It is a process by which pathologies

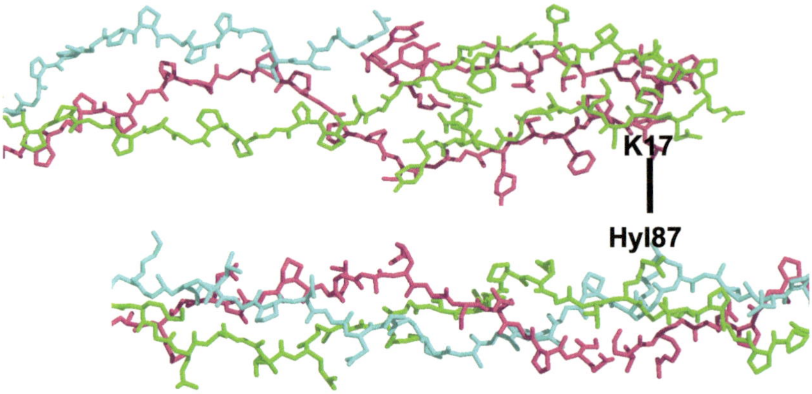

Figure 2. Crosslinks at the C-terminal telopeptide formed as a result of lysyl oxidase activity. A crosslink between lysine (residue 17) on the C-telopeptide is shown with a hydroxyproline (residue 87) on one of the $\alpha1$ chains within the triple helical region.

associated with diabetes progress. The process of glycation involves the generation of a series of intermediates including Schiff bases and Amadori products, and ultimately results in the formation of advanced glycation end-products (AGE). This is a slow process, spread over a span of a few weeks, making biochemical species with relatively long half-lives, such as collagen, prime targets for these reactions [20]. Glycation is a major complication in diabetic patients and is common in both afflicted animal models and humans [21]. The availability of high concentrations of glucose in the blood stream leads to (as yet to be determined) changes in the packing structure of collagen which effects and probably impairs, both the functional and mechanical aspects of collagen's role in the ECM (**Figure 3**). If considerable changes are made to its mechanical properties, those alterations may in turn, induce or further aid in the progression of pathological conditions. Considering the role of fibrillar collagens in the structure of vasculature, as a sole target discussed among several possibilities, the significance of these changes becomes readily apparent. AGEs are heterogeneous and complex groups of compounds, strongly implicated in the pathobiology of diabetes [21]. The initial chemistry behind the formation of AGEs has been known since the early 1900s, although a great deal of important information with regards to the contribution of AGEs in disease has been uncovered in the last 30 years, even while the specific structural sites effected by these changes were yet to be determined.

The process of glycation is a Maillard reaction that can be divided into three broad stages [22]. The first stage of glycation is a result of a nucleophilic attack by the amine group, most commonly the ε-amino group (side chain) in amino acids such as lysine, on the carbonyl (—CHO) group of the sugar in open chain confirmation to form a Schiff base. This is a reversible step [23]. Both pentoses and hexoses can take part in this reaction, although only in open chain conformation which gives the otherwise closed carbonyl group access to react with the amines from the protein. This Schiff base is then rearranged into a relatively more stable Amadori product [22, 24].

The Amadori product formed in the first stage of the glycation reaction breaks down to give rise to several compounds, such as glyoxal, methylglyoxal, and 3-deoxyglucosone and various other smaller intermediates [22, 24, 25].

http://dx.doi.org/10.5772/intechopen.71022

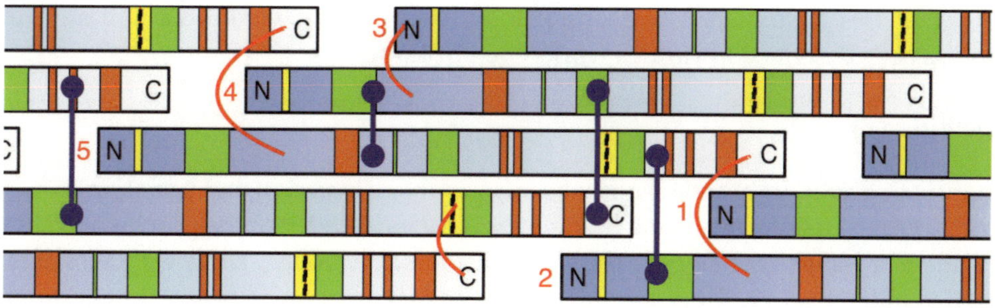

Figure 3. A magnified representation of the microfibrillar arrangement of collagen monomers. Mature enzyme-mediated (lysyl oxidase) crosslinks are marked in orange and possible, candidate, sugar-mediated (glycation) crosslinks are marked in blue. The colored bands in the collagen monomers show ligand binding sites as depicted in **Figure 1**.

The final/late stage of glycation is a result of the highly reactive α-dicarbonyls from the intermediate stage, reacting with amine and other groups from basic amino acids. After further rearrangement of the chemical structures, a large set of potentially varied and stable AGEs are formed which permanently alter the structure of the proteins that participated in the reaction. These steps are represented in a flow diagram in **Figure 4**. For a detailed discussion on the chemical reaction and the intermediates, see above referenced works.

The structural nature of AGEs is varied. Some AGEs, such as carboxymethyllysine (CML) and carboxymethylarginine (CMA), exhibit no special structural characteristics. Some examples of lysine-lysine AGEs include glyoxal-lysine dimer (GOLD), methylglyoxal-lysine dimer (MOLD), while pentosidine, glucosepane, DOGDIC, MODIC, GODIC are lysine-arginine crosslinks [24]. As already mentioned, the overall kinetics of the non-enzymatic process of glycation are rather slow. The availability of reducing sugars, mainly glucose, in open chain form is fairly limited in the human body but this becomes a significant issue for some common human conditions. Poor glucose control, such as that seen in diabetic patients, can lead to formation and deposition of AGEs in various tissues [26]. Metabolic products of glucose, such as glucose-6-phosphate, tend to be faster glycating agents and to react more quickly with

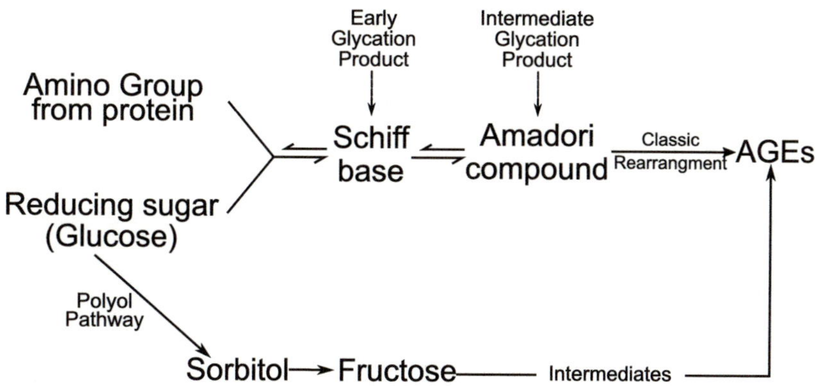

Figure 4. A schematic representation of the process of glycation *in vivo*.

proteins than this sugar in its native form [20, 27]. The pathological effects of glycation of collagens has been described in various works cited in **Table 1**.

1.6. Detection of AGEs

The nature of glycation in collagen has been a point of discussion and scientific inquiry for at least 30 years. Historically, some researchers have indicated that the process and the location of these crosslinks are more or less random and uncontrolled. However, Wess et al. [28] speculated based on their X-ray diffraction studies that a few small peaks in their difference Fourier maps along with a major peak at the lysine 434 in the gap region of the $\alpha 1$ chain, may originate from glycation, and that the process of glycation may be much more specific than it was (at that time) assumed. The ability to find the locations of these changes has further been fundamentally confounded by the rather small electron density changes brought about by the crosslink products, i.e. the AGEs. This makes their detailed study impractical with X-ray diffraction, unless additional X-ray diffraction contrast is provided, i.e. added electron density to sites specific to these glycation events. This study addresses that need.

Boronic acid derivatives have been used to bind with AGEs and other chemical moieties of this class to reproducibly create fluorescent contrast [29–31] in previous experimental work, unrelated to X-ray diffraction. Danslyated boronic acids have been used in determining the concentration of glycation mediated crosslinks in globular and fibrillar proteins and fats have been developed and commonly adopted [32]. These fluorescent contrast agents also happen to have, intrinsic to their nature, some significant electron density. Thus, although their structures are too large to be functionally useful for high-resolution single crystal crystallography, they may be exceedingly useful at lower resolution

Publication and year	Glycation associated pathology
Stitt et al., 2005, 1998	Destabilization of the structure of the vitreous gel within the eye
Albon et al., 1995	Age-related changes in the human lamina cribrosa, and the prevalence of chronic open angle glaucoma (COAG)
Dyer et al., 1993	Changes in mechanical properties – stiffer and more brittle collagen
Brownlee, 2000	Altered cell proliferation, motility, gene expression and response to cytokines
Gugliucci and Bendayan, 1996	Increased activation of cytokine (NF-kB) activity resulting in activation of Matrix Metalloproteinases (MMP) and increased collagen digestion
Murillo et al., 2008	Impaired wound healing as a result of altered periodontal cell behavior
Thomas & Lascelles, 1966	Accumulation of collagen in the endoneurial interstitium leading to impaired neuronal function
Sternberg et al., 1995	End-stage renal disease from glycation of basement membrane in glomeruli
Avendano et al., 1999	Impaired myocardial function from stiffness in collagen in the ventricles

Table 1. Evidence of collagen and AGE related pathologies.

http://dx.doi.org/10.5772/intechopen.71022

(nanometer scale) X-ray diffraction analyses; studies akin to those reported here, where they effectively form isomorphous derivatives, beyond the upper resolution limit of the data collected. Based on these first principles, we present here a method and preliminary results we use to identify the location of AGEs within the fibrillar packing of type I collagen in rat-tail tendons.

2. Materials and methods

All X-ray diffraction experiments were conducted *in situ* on rat-tail tendons (RTT). Rat-tail tendons were carefully dissected out from the tails of healthy adult rats aged 3–6 months. These animals were euthanized, at the time, for other animal studies and the tails were generously donated. The tails were immediately stored at −40°C until the time of dissection for tendons. After dissection, all tendons were stored at 4°C in 10–50 mM Tris Buffered Saline (TBS) (Sigma Aldrich St. Louis MO; Cat no. T5912; 10x stock, diluted as needed). 0.04% (w/v) sodium azide (NaN_3) was added to all storage buffers and sample incubations to prevent microbial contamination.

2.1. Sample setup for glycation

0.2 M solution of glucose was made by slowly mixing the appropriate amount of D-glucose (Sigma Aldrich St. Louis MO; Cat no. 158968) and the serially diluting the stock to make the desired concentrations. All sugar solutions were made in 10 mM TBS. Tendons were deposited in Falcon tubes containing each of the sugar solutions and incubated in a 37°C water bath for 20–40 days.

2.2. Treatment with fluorescent dyes

3-(Dansylamino)phenylboronic acid (DPBA) (Sigma Aldrich (St. Louis MO; Cat No. 30423-) is a fluorescent dye that binds to the AGEs [32]. A 30 μM stock of DPBA was made in methanol. A working standard DPBA solution at 3 μM was made by diluting the using Phosphate Buffered Solution (PBS), at pH 7.4. Non-glycated and glycated tendons were incubated in the working standard solution of DPBA for 20 minutes and then washed in PBS for 2 minutes, prior to XRD data collection.

2.3. Small and wide angle X-ray diffraction

RTT diffraction patterns were obtained at Biophysics Collaborative Access Team (BioCAT) facility at Argonne National Laboratory, Chicago IL. The Mar 165 CCD detector was used to collect all small (1–15 Bragg orders) and wide-angle (3–22 Bragg orders) diffraction patterns. The sample to detector distances for small angle (SAX) and wide angle (WAX) X-ray diffraction were from 1.5 to 2.5 m and from 25 to 70 cm, respectively. Beamline optics were setup to deliver focused X-ray beams at sizes between 50 and 100 μm in width and height (at an average flux of 10^{13} photons/s) for SAX studies. KB mirrors were used to deliver focused beam at 10–60 μm for WAX and micro-WAX studies at a flux of ~1×10^{-12} photon/s. The sample mount was positioned so that the sample was held at the focal point of the beam. The samples were mounted onto the sample compartment, shown in **Figure 5**, immediately before data collection. The sample compartment was mounted on an XY positioner to move the sample into the beam path. Multiple patterns from each sample were collected and the sample was moved, vertically and/or horizontally to limit radiation damage and improve the quality of data. Each point on the tendons was exposed to no more than 1 second to the X-rays. Based on the flux of the available beam, if saturation of pixels was observed on the diffraction pattern, additional attenuator foils were deployed. The same data collection strategies were used to collect data from control, glycated and iodine treated samples.

Amplitudes of each reflection from the diffraction patterns were obtained by azimuthally integrating the data from a sector covering the meridional reflection series (**Figure 6**). Difference Fourier and Patterson analyses was performed using the amplitudes and phases from native type I collagen as described previously [6, 11, 33]. After appropriate scaling, the data from SAX and WAX measurements were combined to improve the resolution of the interpretations.

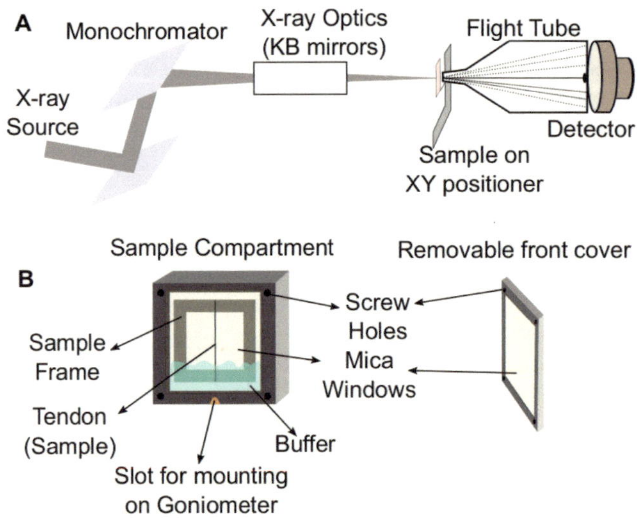

Figure 5. XRD beamline setup and a standard sample cell for fiber XRD data collection. (A) KB mirrors were used to deliver a focused beam at the sample position. The sample compartment was placed on a motorized XY-positioner to move the sample at the time of recording data. An evacuated flight tube is used to deliver the diffracted (scattered) X-rays to the marCCD detector at the end of the tube. The length of the tube can be varied for SAX and WAX experiments. (B) Research grade mica windows (thickness 0.15–0.21 mm) are used to cover the windows on either sides of the chamber to minimize data loss from any diffracted ray absorption from the compartment itself. An inner sample frame is used to tie the tendon onto and then be placed inside the compartment. A cylindrical slot in the bottom enables the entire rig to be mounted onto a goniometer head, which can then be controlled remotely. The buffer at the bottom of the compartment allows for hydration of tendon by capillary action.

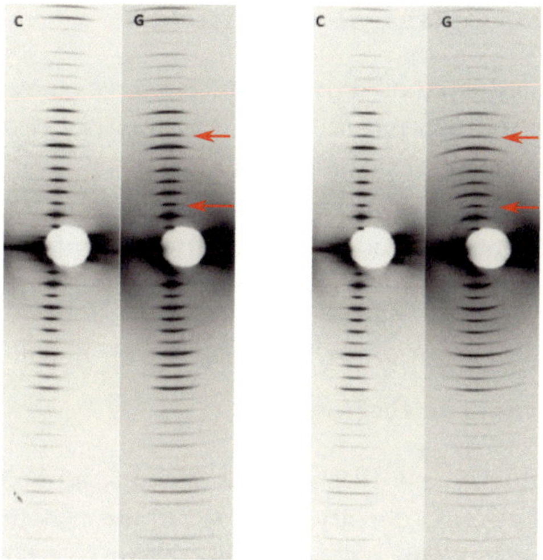

Figure 6. WAX diffraction patterns from non-treated (C), glucose- (G) and treated tendons. Red arrows show orders where an increase in the intensity of reflection was observed.

3. Results and discussion

X-ray diffraction data were interpreted to determine the location of both native crosslinks (those formed normally in non-diseased tissue) and those formed through the process of sugar incubation *in vitro*. As stated before, the latter sugar incubation induced cross-links simulate the *in vivo* process of formation of crosslinks through glycation/normal aging. Difference Fourier/electron density maps, to a resolution of ~35 Å (19 meridional orders) were calculated from these X-ray data. Interpretation of the D-periodic location of existing and newly formed crosslinks was then validated by comparison with difference Patterson maps. This verified the correct scaling of derivative to native data, the isomorphous nature of the derivative data and finally, the correct interpretation of where these crosslinks are from comparison of model Patterson functions based on the locations determined for crosslinks [6, 11, 33]. The peaks in difference Patterson maps are a result of addition of electron densities to specific regions within the unit cell, one of the hallmarks in processing of isomorphous data [9]. The cross verification of the labeling positions determined from difference Fourier maps with the relative periodicity spacing indicated by the difference Patterson maps provides information on the location (and plausible identity) of these crosslinks. This is important, since the Patterson maps derived from the experimental data, are unbiased by experimenter assumptions. Therefore, the close match between difference Patterson functions calculated from observed data and model Patterson functions calculated from the structure factors derived from a D-periodic model of crosslink locations, provides robust verification of the correct determination of the crosslink locations; all the more so, since the model of crosslink location is derived directly from difference Fourier maps.

3.1. Preliminary diffraction experiments with glycation

Data collected from glycated tendons showed minor changes in relative intensities (**Figure 6**). Considering the very small change in electron density brought about by the freshly formed, sugar-mediated crosslinks, this is to be expected. The small changes detected in intensities, however, indicate that there may be isomorphous changes that can be better characterized with increased electron density contrast. Contrast agent that binds (also isomorphously) to the reaction products (AGEs) where used to make these crosslinks more "visible" to X-rays.

3.2. Use of fluorescent dyes to detect crosslinks

As indicated, the use of fluorescent dyes was mandated to add electron density contrast, via isomorphous addition of electron density (contrast) to the crosslinks. Compounds known to bind to AGE-derived crosslinks for collagen were selected as potential candidates for multiple isomorphous replacement (MIR). 3-(Dansylamino) phenylboronic acid (DPBA) has been demonstrated to increase fluorescence signal by binding to the reaction product from glycation of bovine serum albumin (BSA) [32]. This property was used to increase the contrast of crosslinks (native and AGEs) in glycated and non-glycated collagen samples to reflect in the difference Fourier and Patterson maps.

3.3. Difference electron density analysis

As described earlier, difference Fourier and Patterson analysis was performed using the amplitudes of the Bragg reflections extracted from the diffraction patterns of glycated and non-glycated (control) tendons after treatment with DPBA. **Table 2** shows the matrix of difference Fourier and Patterson analyses performed.

The addition of electron densities due to glycation can be observed in the comparison between the difference Fourier maps displayed in **Figure 7**. Difference Patterson maps, generated from these diffraction amplitudes is show in **Figure 8**. The series of added densities can then be

	Non-glycated	Non-glycated + DPBA	Glycated	Glycated + DPBA
Non-glycated	–	Native Crosslinks (with contrast)	Sugar Crosslinks (without contrast)	Native + Sugar Crosslinks (with contrast)
Non-glycated + DPBA		–	Invalid calculation	Sugar Crosslinks (with contrast)
Glycated			–	Native Crosslinks
Glycated + DPBA				–

Table 2. Table showing difference Fourier and Patterson calculated from observed data. A difference Fourier/Patterson is simply put, the difference series calculation (the Fourier or the Patterson) between two data sets of structure functions (amplitudes). The red and blue text represent the respective series of the same colors shown in **Figure 7**.

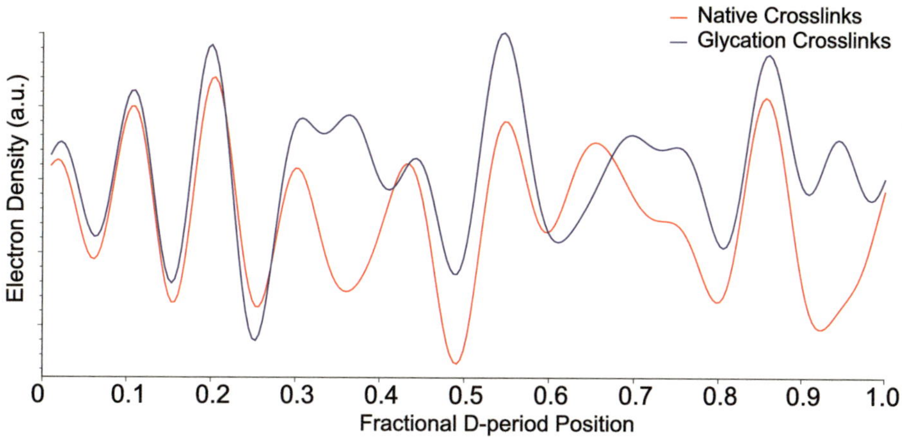

Figure 7. Difference Fourier map from glycated tendons treated with DPBA (blue) and non-glycated (native) tendons treated with DPBA (red) along 1D period represented fractional positions. The increase in electron densities or the introduction of new peaks in the glycated series, in comparison to the non-glycated, are a result of AGE crosslinks formed from the process of glycation.

used to synthesize a model difference Patterson to compare to that obtained from the experimental data to confirm scaling and the application of the right phase information (**Figure 8**). The specific sites of increased electron densities are better demonstrated in **Figure 9**.

3.4. Interpretation of sites of glycation

Lysine (K), arginine (R) and hydroxylysine (U) residues are the principal residues for crosslink (both enzymatic and non-enzymatic) formation. Close proximity of these residues to each

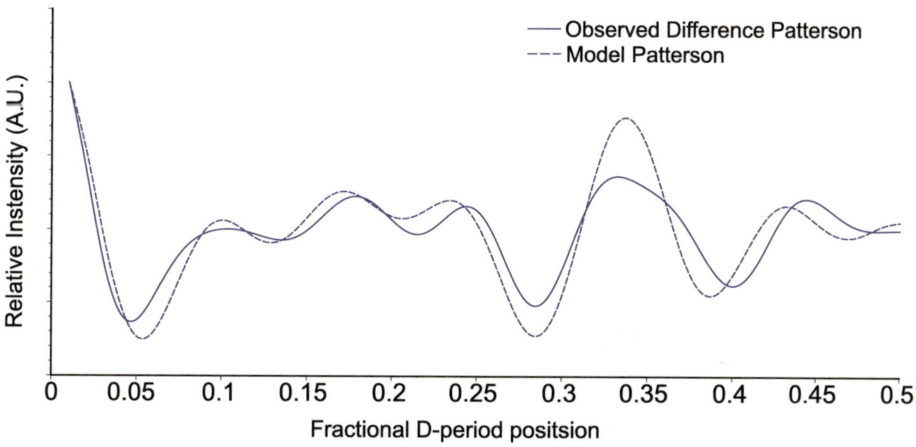

Figure 8. Model vs. observed difference Patterson maps. Difference Patterson map between glycated and non-glycated tendons with DPBA (solid blue line) along 1D period represented in fractional positions. The peak positions show the distances between electron dense regions seen in the difference Fourier map (**Figure 7**). The similarity model Patterson (dashed blue line) confirms that the scaling of derivative amplitudes from the XRD patterns and the application of native phases is appropriate, thus validating the sequence and structure specific interpretations.

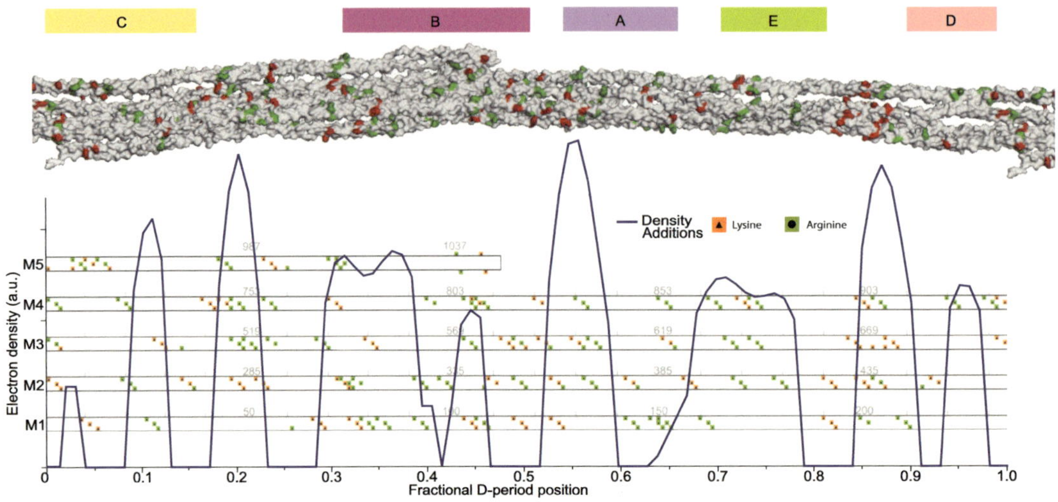

Figure 9. Candidate sites for glycation and regions of increased electron densities from crosslinking within a collagen D-period. A linear arrangement of amino acid sequences is shown with lysines (K) and hydroxylysines (U) marked in orange and arginines (R) residues marked in green. These positions are aligned with a 3D surface rendered model of the collagen microfibril. The linear and 3D representations are also presented in the context of the difference Fourier map (**Figure 7**) with the sites showing regions of increased electron density. An increase in electron density is observed around clusters of candidate residues in both representation, thereby supporting the validity of the results. See the discussion on the amino acid residues in each of these regions and their relevance to physiology and possible pathologies.

other within the three-dimensional structure is also a prerequisite to crosslink formation. Two very distant candidate residues are unlikely to be involved in crosslink formation, when compared to closely located residues, which cannot be reliably determined without reference to the three-dimensional structure. Candidate sites for glycation are spread out throughout the sequence (more specifically, the D-period) of collagen (**Figure 9**). These sites are of particular importance for the formation of mature enzymatic crosslinks between the non-helical telopeptides.

In the N-terminal telopeptide (0.025D) there are clusters of K and U residues which participate in the lysyl oxidase-mediated intermolecular crosslink formations. Furthermore, a series of R residues in this locale can be involved in AGE formation and hyperstabilize the enzymatic formed crosslinks at the N-terminus. Based on our data, these AGEs could potentially compete for the candidate K and U residues involved in the formation of enzymatic crosslinks, thus pathologically changing the nature of these crosslinks. Similarly, our data indicate at the C-terminal telopeptide (0.43D), a cluster of R residues participate in AGE crosslinking [16]. Candidate K and U crosslinking residues on the C-terminal telopeptide are normally the sites for enzymatic-intermolecular crosslinking. Together the N- and the C- terminus enzymatic crosslinks are prominently involved in the stabilization of the packing structure of collagen into fibrils. A possible site for glycation crosslink, adjacent to the N-terminal telopeptides was reported by Sweeney et al. [34].

Glycation-mediated crosslinks and possible sites for the formation of these crosslinks are described here. The results reported here now provide us with a context for interpretation of the 3D structure of type I collagen with appropriate AGE-related crosslinks and study their effects in physiology and disease. For instance, our data indicate the formation of AGE crosslinks near the location of the high affinity integrin and von Willebrand Factor (vWF) sites. This may have great significance towards explaining the poor wound healing properties of diabetic tissues. If these crosslinks obscure these high-affinity cell-binding domains in the ECM, the tissue is likely significantly impacted in its ability to respond to both wound management and the repair of normal wear and tear. Potentially, these data can be used in the design of improved therapies, now that a molecular cause is indicated (see following discussion).

Notable changes are indicated by increased electron density of isomorphous derivatives, evident in difference Fourier maps, along the D-period (67 nm) of type I collagen fibrils in those derivative RTT tissues incubated with high concentrations of glucose. This is indicative of significant structural changes, i.e. crosslinking, and is observed from the experimental data. That these changes are crosslinks is further supported by the presence of crosslink labile amino acid sequences found in the regions of the D-period in which the additional electron density is observed, thereby confirming that the process of glycation is site-specific. The fact that these crosslinks alter the packing structure and functional (as in cell interactive for instance) properties of collagen is indicated by the proximity of these clusters of crosslinks to highly significant functional sites and regions in the collagen packing structure.

The peak indicating additional crosslinking, that appears at 0.1D confirms the glycation site proposed by Sweeney et al. [34], with a crosslink between monomers 3 and 4 of the microfibril. Additionally, our data indicate possible AGE crosslinking between monomers 1, 2 and 4. The peak at around 0.22D is observed, for which a cluster of principal sites is responsible. This location is of particular importance as it is located around the region where monomer 4 of one microfibril interacts with another neighboring collagen. The additional crosslinks around this location might hyper-stabilize and/or obscure the resulting interaction. This region is also closely placed with one of the $\alpha_1\beta_1$ and $\alpha_2\beta_1$ integrin binding sites (GFOGER) [35]. It is worth noting that one of the residues in this recognition site is an arginine, which might directly be involved in the AGE crosslink. This may well alter integrin-mediated cell adhesion through either direct interference (steric hindrance of integrin binding) or by obstructing/delaying the limited proteolytic activity of neighboring collagen molecules needed before integrin binds at this site.

The crosslinking electron density peaks at 0.3D, 0.38D and 0.46D surround a cluster of key cell interaction sites. For example, these AGEs might be significant with respect to the fibronectin and MMP binding sites. Alterations in this physiological phenomenon can severely change the accessibility and deposition of collagen in wound healing. Also within close proximity of this crosslink are several RGD sequences, which are principal recognition sites for 8 of the 24 integrins and a possible explanation for advanced rate of aging and poor wound healing ability in diabetic patients [36]. The peaks indicating crosslinks at 0.38D and the 0.46D are in the region of the $(GPO)_5$ site, which is a potent site for platelet binding and the process of clotting [4]. This could constitute part of a plausible explanation for the suppressed blood clotting in diabetic patients. This is also of great importance in the development of successful surgical implantation materials, such a heart valves and capillaries for anastomoses where the ability of the implant material to support clotting is vital. Furthermore, a crosslink at 0.46D (gap-overlap interface) can interfere with the fibrillogenesis control sequence (FCS) in this region. This site controls the formation of fibrils by forming mature crosslinks to stabilize the molecular packing of collagen monomers into fibrils and fibers. The crosslinks at the gap-overlap interface could also alter the packing in such a manner that the MMP interaction site becomes more accessible, thereby leading to increased proteolysis. Our data also suggest that the crosslink at 0.46D can also interfere with the GMOGER site in this region. This has been identified as an important site for integrin-mediated cell binding.

The 0.55D crosslink sites appears immediately adjacent to the C-terminal telopeptides and at the beginning of the gap region. At this location, monomer 2 of one microfibril can form a crosslink with neighboring collagen monomers to stabilize the packing structure. The significance of 0.55D crosslinks are not obvious until it is considered in the context of the 0.38, 0.46 and the aforementioned 0.55D clusters of crosslinks. Together, they surround the enzymatic cleavage-site of collagenase for fibrillar collagen around 0.44D on monomer 4. If there were significant glycation induced crosslinking on either side of the collagenase cleavage site, then it is possible that even if the enzyme is able to access its substrates-site, which will not be a given, it might have little or no impact on the proteolysis of fibrillar collagen as the microfibril and fibril will remain intact due to the pathological crosslinks bridging the enzymatic break in the collagen molecule. In other words, extensive AGE crosslinking can be seen in this molecular model, to prevent or hinder wound healing and removal of damaged collagenous tissue, which coincides with clinical observations also [21, 37–39].

The peaks indicating crosslinks at the 0.7–0.77D (and a separate peak at 0.88D) are due to the high concentration of principal sites of glycation in these regions. In physiology, this region is responsible for the binding of small leucine-rich repeat proteoglycans (sLRRP PGs), which, using their long glycosaminoglycan chains (GAG) help in bundling fibrils into stable structures [5]. The crosslinks formed at these sites could result in inherent detachment of the PGs leading to disintegration of fibrils. Similar disintegration of fibrils in type II collagen in human articular cartilage, as a result of treatment with the biglycan (PG) antibody has been exhibited as a plausible mechanism for rheumatoid arthritis [33]. Similar conclusions can be drawn with the disintegration of type I collagen fibrils resulting in disease. This region is also the principal recognition site for von Willebrand factor binding and crosslinks here could lead to faulty coagulation in diabetics. These crosslinks are a possible explanation to how the process of thrombosis is affected in diabetic patients [40]. The 0.95D crosslink forms a link between the N-terminal of the neighboring collagen molecule. This region is the principal binding region for amyloid precursor protein (APP) and is implicated in the pathology of Alzheimer's disease.

Although all potentially functionally significant, we draw attention again to the significance of the additional glycation induced crosslinking that is indicated to occur at 0.38, 0.46 and

0.55D. As these locations surround the collagenase cleavage-site, these crosslinks could be implicated to interfere with the normal and healthy removal and repair of damaged collagenous tissues. Crosslinks effect these processes, firstly by limiting the availability of the collagen's cleavage site and interaction region at 0.44D; secondly, by hyperstabilizing the packing structure, even after the enzymatic cleavage occurs, thereby defeating the release of unhealthy collagen fibrils. The 0.7–0.77D region is a second area indicating significant glycation induced crosslinking, which directly interferes with and/or hypostabilizes principal matrix-matrix association regions (via proteoglycan mediated inter-fibrillar crosslinking GAG chains) and notable, the von Willebrand factor interaction site in the fibrillar collagen's gap region.

4. Conclusions

There is a strong correlation between the availability of sites for glycation, namely lysine, arginine and hydroxylysine residues, their placement in the microfibrillar arrangement and the electron density peaks indicating crosslinking in sugar incubated tendons (**Figure 9**). The regions of increased electron density, as a result of glycation induced crosslinks, directly correspond with candidate amino acid crosslinking sites along the collagen microfibril. This data appears to experimentally confirm several previously estimated locations for crosslinking [34, 41]. The significance of these data to our understanding of the fundamental structure of connective tissues and the effects of AGE induced cross-links are immediately apparent as discussed above and as follows.

The structure and packing of type I collagen have been objects of study for many decades, with the resolution of packing coming to a better level of understanding. There are various structural features that contribute to the final fibrillar packing, resulting in the classic 67 nm repeating D-period. Crosslinking is one of the most important features to stabilize their packing into fibrils and in turn, the organization of these fibrils into fibers and the bases of tissues and organogenesis. The three-dimensional structure and its derived model for collagen, obtained from high-resolution X-ray diffraction studies [2, 7, 16] accounts for enzyme-mediated crosslinks, formed as a result of post-translational modification. However, molecular dynamics simulations of the packing of type I collagen predict a 11–19% shrinkage of the 67 nm D-period – which is not what is observed in nature! Therefore, there are additional parameters required for molecular dynamic simulations to resemble the observed natural structure. Some of this can be attributed to the formation of AGEs as a result of normal physiological aging and/or sugar-mediated crosslinking [42]. The data presented here, give us insight into the location of these crosslinks including which amino acid residues are involved and can be used to refine the packing structure and associated simulations of type I collagen. These simulations can be used to study plausible collagen related disease mechanisms, to support experimental and clinical data and aid with the development of appropriate therapeutics.

Author details

Rama Sashank Madhurapantula[1] and Joseph P.R.O. Orgel[2*]

*Address all correspondence to: orgel@iit.edu

1 Department of Biology, Pritzker Institute of Biomedical Science and Engineering, Illinois Institute of Technology, Chicago IL, USA

2 Departments of Biology, Physics and Biomedical Engineering, Pritzker Institute of Biomedical Science and Engineering, Illinois Institute of Technology, Chicago IL, USA

References

[1] Orgel JPR, Miller A, Irving TC, Fischetti RF, Hammersley AP, Wess TJ. The in situ supermolecular structure of type I collagen. Structure. 2001;**9**(11):1061-1069. DOI: 10.1016/S0969-2126(01)00669-4

[2] Orgel JPRO, Irving TC, Miller A, Wess TJ. Microfibrillar structure of type I collagen in situ. Proceedings of the National Academy of Sciences. 2006;**103**(24):9001-9005. DOI: 10.1073/pnas.0502718103

[3] Orgel JPRO, Irving TC. Advances in fiber diffraction of macromolecular assembles. In: Meyers RA, editor. Encyclopedia of Analytical Chemistry. Chichester, UK: John Wiley & Sons, Ltd; 2014. pp. 1-26

[4] Orgel JPRO, San Antonio JD, Antipova O. Molecular and structural mapping of collagen fibril interactions. Connective Tissue Research. 2011;**52**(1):2-17. DOI: 10.3109/03008207.2010.511353

[5] Orgel JPRO, Eid A, Antipova O, Bella J, Scott JE. Decorin core protein (decoron) shape complements collagen fibril surface structure and mediates its binding. PLoS One. 2009;**4**(9): e7028. DOI: 10.1371/journal.pone.0007028

[6] Antipova O, Orgel JPRO. In situ D-periodic molecular structure of type II collagen. Journal of Biological Chemistry. 2010;**285**(10):7087-7096. DOI: 10.1074/jbc.M109.060400

[7] Orgel JP. The molecular structure of collagen [thesis]. University of Stirling; 2000

[8] Madhurapantula RS. Studies on connective and neurological tissues in relation to disease [thesis]. Ann Arbor: Illinois Institute of Technology; 2015

[9] Blundell TL, Johnson LN. Protein Crystallography 1st ed. United States: Elsevier Science Publishing Co Inc; 1976. ISBN: 9780121083502

[10] Orgel JPRO, Irving TC. Advances in fiber diffraction of macromolecular assembles. In: Encyclopedia of Analytical Chemistry. John Wiley & Sons, Ltd; 2006. DOI: 10.1002/9780470027318.a9420

[11] Orgel JPRO, Persikov AV, Antipova O. Variation in the helical structure of native collagen. PLoS One. 2014;**9**(2):e89519. DOI: 10.1371/journal.pone.0089519

[12] Eyre DR, Paz MA, Gallop PM. Cross-linking in collagen and elastin. Annual Review of Biochemistry. 1984;**53**(1):717-748. DOI: 10.1146/annurev.bi.53.070184.003441

[13] Henkel W, Glanville RW. Covalent crosslinking between molecules of type I and type III collagen. European Journal of Biochemistry. 1982;**122**(1):205-213. DOI: 10.1111/j.1432-1033.1982.tb05868.x

[14] Last JA, Armstrong LG, Reiser KM. Biosynthesis of collagen crosslinks. International Journal of Biochemistry. 1990;**22**(6):559-564. DOI: 10.1016/0020-711X(90)90031-W

[15] Sims TJ, Avery NC, Bailey AJ. Quantitative Determination of Collagen Crosslinks. In: Streuli CH, Grant ME. (eds) Extracellular Matrix Protocols. Methods in Molecular Biology, vol 139. Humana Press; 2000. p. 11-26. DOI: 10.1385/1-59259-063-2:11

[16] Orgel JP, Wess TJ, Miller A. The in situ conformation and axial location of the intermolecular cross-linked non-helical telopeptides of type I collagen. Structure. 2000;**8**(2):137-142. DOI: 10.1016/S0969-2126(00)00089-7

[17] Mitome J, Yamamoto H, Saito M, Yokoyama K, Marumo K, Hosoya T. Nonenzymatic cross-linking pentosidine increase in bone collagen and are associated with disorders of bone mineralization in dialysis patients. Calcified Tissue International. 2011;**88**(6):521-529. DOI: 10.1007/s00223-011-9488-y

[18] Tomkins O, Garzozi HJ. Collagen cross-linking: Strengthening the unstable cornea. Clinical Ophthalmology (Auckland, N.Z.). 2008;**2**(4):863-867

[19] Alhayek A, Lu P-R. Corneal collagen crosslinking in keratoconus and other eye disease. International Journal of Ophthalmology. 2015;**8**(2):407-418. DOI: 10.3980/j.issn.2222-3959.2015.02.35

[20] Singh R, Barden A, Mori T, Beilin L. Advanced glycation end-products: A review. Diabetologia. 2001;**44**(2):129-146. DOI: 10.1007/s001250051591

[21] Peppa M, Uribarri J, Vlassara H. Glucose, advanced glycation end products, and diabetes complications: What is new and what works. Clinical Diabetes. 2003;**21**(4):186-187. DOI: 10.2337/diaclin.21.4.186

[22] Priego Capote F, Sanchez J-C. Strategies for proteomic analysis of non-enzymatically glycated proteins. Mass Spectrometry Reviews. 2009;**28**(1):135-146. DOI: 10.1002/mas.20187

[23] Brownlee M, Vlassara H, Cerami A. Nonenzymatic glycosylation products on collagen covalently trap low-density lipoprotein. Diabetes. 1985;**34**(9):938-941

[24] Cho S-J, Roman G, Yeboah F, Konishi Y. The road to advanced glycation end products: A mechanistic perspective. Current Medicinal Chemistry. 2007;**14**(15):1653-1671. DOI: 10.2174/092986707780830989

[25] Thornalley PJ, Langborg A, Minhas HS. Formation of glyoxal, methylglyoxal and 3-deoxyglucosone in the glycation of proteins by glucose. The Biochemical Journal. 1999; **344**(Pt 1):109-116

[26] Snedeker JG, Gautieri A. The role of collagen crosslinks in ageing and diabetes - The good, the bad, and the ugly. Muscles, Ligaments and Tendons Journal. 2014;**4**(3):303-308

[27] Pischetsrieder M. Chemistry of glucose and biochemical pathways of biological interest. Peritoneal Dialysis International: Journal of the International Society for Peritoneal Dialysis. 2000;**20**(Suppl 2):S26-S30

[28] Wess TJ, Miller A, Bradshaw JP. Cross-linkage sites in type I collagen fibrils studied by neutron diffraction. Journal of Molecular Biology. 1990;**213**(1):1-5. DOI: 10.1016/S0022-2836(05)80115-9

[29] Mallia AK, Hermanson GT, Krohn RI, Fujimoto EK, Smith PK. Preparation and use of a boronic acid affinity support for separation and quantitation of glycosylated hemoglobins. Analytical Letters. 1981;**14**(8):649-661. DOI: 10.1080/00032718108055476

[30] Hayashi Y, Makino M. Fluorometric measurement of glycosylated albumin in human serum. Clinica Chimica Acta. 1985;**149**(1):13-19

[31] Guan Y, Zhang Y. Boronic acid-containing hydrogels: Synthesis and their applications. Chemical Society Reviews. 2013;**42**(20):8106. DOI: 10.1039/c3cs60152h

[32] Sattarahmady N, Moosavi-Movahedi AA, Ahmad F, et al. Formation of the molten globule-like state during prolonged glycation of human serum albumin. Biochimica et Biophysica Acta. 2007;**1770**(6):933-942. DOI: 10.1016/j.bbagen.2007.02.001

[33] Antipova O, Orgel JPRO. Non-enzymatic decomposition of collagen fibers by a biglycan antibody and a plausible mechanism for rheumatoid arthritis. PLoS One. 2012; **7**(3):e32241. DOI: 10.1371/journal.pone.0032241

[34] Sweeney SM, Orgel JP, Fertala A, et al. Candidate cell and matrix interaction domains on the collagen fibril, the predominant protein of vertebrates. Journal of Biological Chemistry. 2008;**283**(30):21187-21197. DOI: 10.1074/jbc.M709319200

[35] Knight CG, Morton LF, Peachey AR, Tuckwell DS, Farndale RW, Barnes MJ. The collagen-binding A-domains of integrins $\alpha 1\beta 1$ and $\alpha 2\beta 1$ recognize the same specific amino acid sequence, GFOGER, in native (triple-helical) collagens. Journal of Biological Chemistry. 2000;**275**(1):35-40. DOI: 10.1074/jbc.275.1.35

[36] Kim JP, Zhang K, Kramer RH, Schall TJ, Woodley DT. Integrin receptors and RGD sequences in human keratinocyte migration: Unique anti-migratory function of alpha 3 beta 1 epiligrin receptor. The Journal of Investigative Dermatology. 1992;**98**(5):764-770

[37] Hennessey PJ, Ford EG, Black CT, Andrassy RJ. Wound collagenase activity correlates directly with collagen glycosylation in diabetic rats. Journal of Pediatric Surgery. 1990;**25**(1):75-78. DOI: 10.1016/S0022-3468(05)80167-8

[38] Peppa M, Stavroulakis P, Raptis SA. Advanced glycoxidation products and impaired diabetic wound healing. Wound Repair and Regeneration. 2009;**17**(4):461-472. DOI: 10.1111/j.1524-475X.2009.00518.x

[39] Peppa M, Brem H, Ehrlich P, et al. Adverse effects of dietary glycotoxins on wound healing in genetically diabetic mice. Diabetes. 2003;**52**(11):2805-2813. DOI: 10.2337/diabetes.52.11.2805

[40] Kessler L, Wiesel ML, Attali P, Mossard JM, Cazenave JP, Pinget M. Von Willebrand factor in diabetic angiopathy. Diabetes & Metabolism. 1998;**24**(4):327-336

[41] Gautieri A, Redaelli A, Buehler MJ, Vesentini S. Age- and diabetes-related nonenzymatic crosslinks in collagen fibrils: Candidate amino acids involved in Advanced Glycation End-products. Matrix Biology. 2014;**34**:89-95. DOI: 10.1016/j.matbio.2013.09.004

[42] Varma S, Botlani M, Hammond JR, Scott HL, Orgel JPRO, Schieber JD. Effect of intrinsic and extrinsic factors on the simulated D-band length of type I collagen. Proteins: Structure, Function, and Bioinformatics. 2015;**83**:1800-1812. DOI: 10.1002/prot.24864

Ion Beam, Synchrotron Radiation, and Related Techniques in Biomedicine: Elemental Profiling of Hair

Karen J. Cloete

Additional information is available at the end of the chapter

http://dx.doi.org/10.5772/intechopen.72076

Abstract

Elements play an imperative role in the physiological and metabolic processes of the human body. When elemental levels deviate from physiologically accepted levels due to for example poor nutrition, the body's intricate elemental and metabolic balance is disturbed. Over time, disease may develop as a result of elemental dyshomeostasis or alternatively, disease may trigger elemental dyshomeostasis as an adaptive metabolic response to an unhealthy environment. There is now a growing interest in screening human tissue to identify and quantify elemental changes as biomarkers of disease or alternatively, as outcomes of disease. The unique properties of human hair brand it the ideal substrate for the quantitative identification of elements in the body. Hair bioaccumulates elements, provides a historical overview of elemental status depending on length, and is easy and economical to sample and store. The fundamental outcome and application of hair elemental screening, however, are strongly influenced by a range of factors, including choice of analytical method. This chapter will provide a background summary of ion beam and synchrotron radiation techniques and its diverse applications for unraveling the elemental signature of hair in various fields.

Keywords: biomedicine, elemental screening, hair, ion beam analysis, synchrotron radiation

1. Introduction

The human body harbors a plethora of mineral compounds that form the lifeline of our multifaceted biological system [1, 2]. To illustrate, major and minor elements play a critical role in metabolic pathways and physiological processes of the human body. However, when toxic elements or other xenobiotic compounds from the occupational or natural environment enter the body, elemental dyshomeostasis ensues that adversely affects the aforementioned processes. Poor nutrition is another contributing factor to elemental dyshomeostasis that results in a modified metabolism.

Besides its effect on metabolic processes, elemental dyshomeostasis in the body may also indirectly accelerate the germination of diseases such as neurological disorders and cancers that in turn may indirectly alter metabolic processes and elemental levels within tissues. A growing number of studies now highlight the complex association between elemental dyshomeostasis and biological disorders [3, 4]. Several reports also highlight the importance of monitoring elemental levels as a measure to treat elemental imbalances and prevent the onset of disease or alternatively, as an outcome of disease [5]. Applying such disease intervention through biological tissue elemental screening is however complicated and requires a more in-depth analysis of the full elemental signature of human tissues in the pre- and postdisease stage.

Elemental profiling of biological tissues now finds diverse applications in the fields of biomedicine, pharmacology, toxicology, and forensic science [6–9]. Of the biological tissues used in elemental profiling studies, hair is gaining increasing popularity for quantitative profiling of elements in the body. Since the elemental content in human hair fibers is generally less than 1%, accurate and sensitive analytical techniques are a necessity for studies focused on quantitative elemental profiling in hair fibers [10].

Studies related to hair elemental screening has mostly relied on conventional analytical techniques such as atomic absorption spectroscopy, inductively coupled plasma mass spectrometry, and inductively coupled plasma atomic emission spectrometry [11, 12]. Sample preparation for these aforementioned techniques presents a major pitfall in that elements may be lost or contaminants introduced during chemical processing of samples. A proposed extraction method may also yield high recoveries for some elements and low recoveries for others [13]. In essence, sample processing for these techniques is destructive in that samples from various donors have to be pooled and chemically processed that effectively destroys historical and spatial information of elements preserved within the length of a single hair strand [14]. Besides sample preparation, the capabilities and limitations of the testing instrument also warrant careful scrutiny. For example, some techniques may only reveal the presence of specific elements due to the instability of the analyte under imperfect experimental conditions or the low sensitivity of the technique [15, 16].

One of the most versatile and sensitive analytical techniques for elemental microanalysis of unprocessed biopsy tissues in a manifold of multidisciplinary fields fall under the umbrella of ion beam techniques [17, 18]. The continual development of the components of the beam setup such as the beam optics and scanning systems, as well as the detection and data acquisition devices, has now firmly placed ion beam techniques at the forefront of elemental analysis research in the biomedical and other fields involving hair elemental screening [19]. Besides ion beam techniques, synchrotron radiation techniques are also becoming increasingly popular in biomedical research related to hair elemental analysis.

This chapter aims to provide an overview of why hair is a popular testing substrate for elemental screening, how elements are incorporated into hair, how to prepare hair for ion beam

http://dx.doi.org/10.5772/intechopen.72076

and synchrotron radiation analysis, and an overview of the versatile applications of ion beam and synchrotron radiation techniques in hair elemental profiling in biomedical studies.

2. Why hair?

Hair is a metabolic end product that incorporates minerals from the blood supply into its keratinous matrix during the growth process [14, 20]. Over time, a temporal profile of normal or abnormal metabolic activity or alternatively, exposure to xenobiotics such as heavy metals is created. Depending on the length of the hair shaft, segmental analysis may then reveal a pattern of exposure that enables a retrospective screening of short term or chronic exposure to chemical compounds [14, 21]. Other unique characteristics of hair include its ability to store higher concentrations of minerals than blood or urine for a number of years [22]. Furthermore, mineral levels in hair do not fluctuate in response to changing physiological and/or environmental conditions such as in blood or urine [23]. Most importantly, hair samples from regions such as the scalp can be collected noninvasively and under close supervision, minimizing the risk of manipulation and cross-contamination, and is easy to handle, transport, and store [14].

Hair has become a popular tool for screening and quantifying for example elemental changes within the body [14, 24]. Alternatively, hair testing may find important applications in clinical medicine such as to (1) determine medicinal or chronic doping, (2) confirm gestational drug use, or (3) assess exposure to toxins and pollutants in the workplace or environment [25, 26]. Hair analysis has also become popular in forensic science to (1) assess drug use history, (2) criminal liability of drug users, and (3) intentional or unintentional poisoning in postmortem toxicology [27–29]. Unfortunately though, there is a lack of reports describing the mineral profile of intact, unprocessed hair in its natural physical and chemical state [14]. Fundamental information on the spatial distribution of elements within hair tissues is also lacking, which is particularly important when distinguishing for example exposure to ingested xenobiotics originating from blood feeding the inner hair tissues or exposure to environmental pollutants that accumulate in the outer hair tissues. Furthermore, how elements and drugs are absorbed and incorporated into hair, either biogenically or diagenetically, are also still poorly understood. These unanswered questions may, however, be probed by techniques such as ion beam analysis and synchrotron radiation that allows quantitative elemental spatial data that may also assist in understanding hair elemental uptake mechanisms.

3. Hair elemental uptake mechanisms and factors that affect hair elemental levels

Understanding hair elemental uptake or incorporation mechanisms and the factors that may influence elemental levels in hair is crucial before correctly interpreting ion beam analysis and synchrotron radiation data [14]. Scientists have proposed three models explaining elemental uptake in hair [30]. The first model proposes that elements may either actively or passively diffuse into the hair shaft from the bloodstream feeding the dermal papilla cells. The second model proposes that elements may diffuse from sweat or other excretions into the hair shaft. Alternatively, powders and vapors may also diffuse into the hair shaft, as proposed by the third model.

Besides the aforementioned routes of entry, the incorporation of elements and their levels in hair may be significantly influenced by a range of other variables. These may include the dose of the chemical exposed to and the origin of the chemical [31]. Hair is often treated with a range of cosmetics such as mineral-based dyes, paints, and bleach for attractive appeal, while various shampoos such as antidandruff formulations that contain zinc and selenium are also frequently used for hygiene purposes [14, 32]. When hair is wetted before applying these formulations, the hair fiber swells and the cuticle cells lift [32], causing the hair fiber to be more permeable to cosmetic agents that are known to contain a variety of chemicals as well as tap water that may contain calcium and lead from plumbing. Bleaching may further increase the permeability of hair in that it destroys hair disulfide bonds that cause hair to be more vulnerable to aqueous solutions containing a variety of chemicals.

Other than cosmetic agents and washing that contributes to endogenous hair elemental levels, heavy metals deposited on the hair surface from atmospheric dust and pollution may also contribute to both exogenous and endogenous hair elemental levels [14, 33]. Interestingly, it has been shown in miners that particulates from mined metals may be deposited on the hair surface [34]. The surface contains distinct regions of varying chemical composition and is a unique site with varying binding affinities for different metals [31, 35]. These binding affinities may, however, be affected by hair acidity and hair cosmetic treatments that add negatively and positively charged ionizable groups to hair. Binding of metals to hair may also alternatively be influenced by the levels of eu- and pheomelanin polyanionic polymers. These compounds have been shown to bind metal cations in vivo or in vitro via electrostatic forces [14, 35].

Once elements are incorporated into hair, various other factors may affect their levels in hair. UV exposure, aging, nutritional deficiencies, morbidity, medication use, and acute metal poisoning has been described to cause natural deterioration and hair structural changes that affect the ability of hair to retain elements [14, 36, 37]. Specifically in disease, the levels of elements in the body may not be optimal and indirectly affect hair growth and hence hair structure. Besides the aforementioned variables, gender and age, ethnicity, and genetic polymorphisms may also influence the structure of hair and hence its ability to retain elements [14, 38–40]. For example with ethnicity, the structure, shape, and melanin levels of hair from different races may affect permeability and hence elemental levels within hair [14, 41]. Ethnicity may also strongly link with cultural habits related to washing and hair treatment regimens, as well as dietary differences that indirectly affect hair elemental levels [42].

4. Hair sampling and preparation

Before hair samples are sourced, it is important to first obtain ethical clearance for working with human tissues. Another important consideration is to obtain informed consent from participants who will be donating their hair. Ethical approval from an institutional ethical board is a necessity. Research institutions usually provide comprehensive information on application and approval processes as well as submission deadlines on their webpages related to the ethical clearance of studies. Ethical clearance of studies involving human tissue may take up to 1 month, depending on the institutional review times. Alternatively, one may also request an expedited review in the cover letter accompanying the submission, since research involving human hair constitutes minimal risk research.

When sampling human hair, it is important to consider the location of sampling. Although scalp hair may be exposed to external contaminants, scalp hair is often the preferred sampling choice [43]. Pubic hair may present the ideal alternative to scalp hair as it is less exposed to contaminants [42]. However, pubic hair differs in morphological and structural properties from scalp hair that may impact on elemental uptake mechanisms. Furthermore, there is a large possibility of participants manipulating samples when not supervised during sampling.

The hair should be sampled as close to the skin as possible with stainless steel or Teflon-coated scissors. Hair may also be plucked so as to extract the hair bulb that may be less exposed to contaminations. Furthermore, the hair bulb is the most active metabolic area of hair and may be particularly interesting to study [44]. For retrospective analyses, it is recommended that long hair be sampled as segmental analyses may reveal a pattern of exposure that enables a retrospective screening of short-term or chronic exposure to chemicals [45, 46]. The hair should not be handled with plastic gloves as polydimethylsiloxane contaminants may be transferred to the hair. These compounds negatively influence data retrieved using certain analytical applications such as scanning ion mass spectrometry (SIMS) [47]. Once the hair samples have been properly processed, it may be stored in paper envelopes.

To produce credible results, hair that is free of surface contaminants should be analyzed. To eliminate surface contaminants such as particulates from pollution, smoking, chemical powders, and metals from mining practices, most studies wash their hair samples [14]. With that said, the question of how specific external contaminants contribute to endogenous content still remains to be more rigorously investigated. Furthermore, if certain external contaminants were to become embedded in hair, distinguishing contribution from endogenous versus exogenous elements then becomes very challenging, rendering washing of hair in essence a negligible step. Nevertheless, specific fields such as drug analysis for forensic purposes place strict emphasis on hair washing protocols as drugs in the form of powder or vapor has been shown to contaminate hair and result in false positives [48]. Alternatively, in pollution or toxicology studies, washing hair samples may simply not be necessary since determining the kind of matter deposited on hair and possibly entering the body is important.

A multitude of washing methods has been described in the literature, of which the most popular is described below [14]. The most popular of these methods include washing hair in an ultrasonic bath with acetone, deionized water, and 0.5% Triton X-100 solution. This process is followed by washing with ultrapure water and air drying. The International Atomic and Energy Agency, however, proposes washing hair with a nonpolar solvent such as double-distilled acetone followed by washings in a polar solvent such as deionized water and acetone. Each washing step should be performed for 10 minutes. Authors may also adapt their hair washing protocol to be most suitable and efficient for their specific type of analysis.

There is, however, a number of concerns related to hair washing protocols. First, no standard washing procedure is currently available in the literature that unknowingly introduces variation in hair elemental data and that has been confirmed after testing different washing protocols [49]. Second, authors may not always provide sufficient information on their washing protocol such as whether samples were sonicated and the specific solvents employed. Thirdly, insufficient washing methods may not remove surface contaminants, while washing methods that are too abrasive may cause damage to the hair sample and cause elements to either leach or diffuse into the hair, which often further complicates the assumption of elements either being of biogenic or diagenetic origin. Alternatively, no contamination may be present then representing [14]. In essence, a standard washing procedure that preserves internal elemental content and washes away external contaminants remains an urgent requirement.

Besides washing protocols for distinguishing between endogenous and exogenous hair elemental content, ion beam techniques have shown great promise for spatial distribution analysis of elements in hair tissues [14, 50]. When preparing hair for ion beam analysis, the hair may either be washed or left untreated. However, exposing the internal hair tissues is critical to assist with spatial distribution mapping of tissues and for interpretation of elemental maps. Longitudinal sectioning of hair may be employed to expose the internal tissues of the hair that facilitates distinguishing elements present at the borders of the hair that may possibly link to contamination or alternatively, biogenic elements occurring in the hair medulla [51]. For this purpose, the hair can be sectioned using a stainless steel metal plate (60 × 110 mm) manufactured with 5–mm-wide grooves of depths ranging between 20 and 80 μm [52]. A single hair is laid in a groove of fitting depth, secured, and sectioned longitudinally with a stainless steel razor blade (**Figure 1**). The sectioned hair may be mounted on a silicon wafer with carbon tape (**Figure 2**) in the case of SIMS at MeV energies [53]. Alternatively, the hair

Figure 1. Light micrograph of a longitudinally sectioned scalp hair fiber (unpublished data).

http://dx.doi.org/10.5772/intechopen.72076

may also be mounted intact in special holders, allowing analyses along the length of a hair. Longitudinal sections of hair are particularly useful for the retrospective assessment of exposure to chemicals.

Some studies also cross section hair and analyze these cross sections with an ion beam technique termed microproton-induced X-ray emission spectrometry [50]. Cross sectioning of hair tissues may be performed using cryosectioning [54] in which hair fibers are fixed to a small piece of adhesive tape, which is embedded in 2% carboxymethyl cellulose and flash frozen in liquid nitrogen for 2 minutes. The frozen specimens are subsequently sectioned using a cryostat set at a temperature of −20°C. Cryosections of approximately 5 μm thickness can be prepared using a cryostat microtome (**Figure 3**). Data from a cross section of hair may well complement data from a longitudinal section in that more information becomes available on the elemental distribution in an area of hair containing different hair tissues.

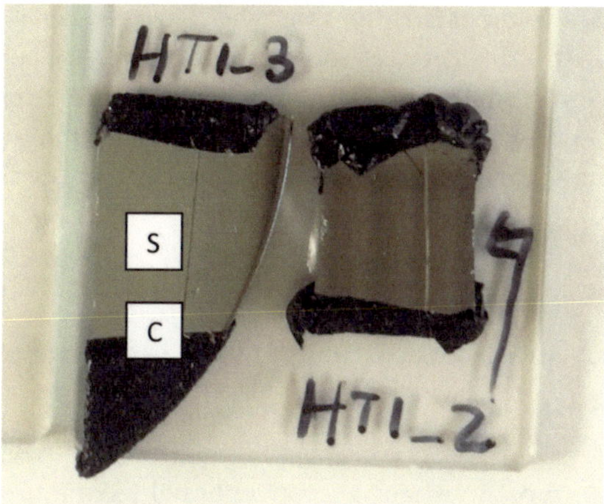

Figure 2. Longitudinally sectioned scalp hair strands secured on silicon wafers (S) with carbon tape (C). The incised surface should be positioned to face the primary ion beam.

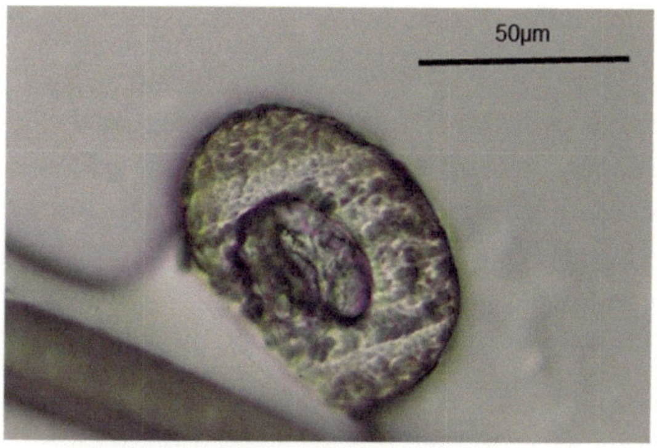

Figure 3. Light micrograph of an animal hair fiber sectioned under cryoconditions (unpublished data).

Such analysis may also effectively aid in determining elements distributed at the root sheath or cuticle cells and medulla of the hair and hence distinguish external contaminants from biogenic elements.

The aforementioned sectioning techniques may only be relevant when screening small sample sizes. However, for population studies, it is more advisable to analyze hair samples from a large number of participants that have been homogenized and pelleted [55]. This approach, however, has several disadvantages. Sampling requires a high number of participants, retrospective analysis will not be possible, information from specific individuals is lost as hair samples are homogenized, and contaminants may be introduced during sample processing. Sampling processing involving bulk processing is, however, fairly simple and traditionally involves homogenizing the hair samples into a powder using the brittle fracture technique. Approximately 2 g of hair are homogenized in a Teflon container and ball, which has been cooled with liquid nitrogen for 3 minutes. The container containing the hair and Teflon ball is vibrated for 2 minutes at 3000 cycles per min using a "micro-dismembrator." The longer the procedure is repeated, the finer the hair powder obtained. The fine powder should be carefully mixed, stored at room temperature, and pelleted before analysis. Graphite (1% pure reactor grade) may also be added to the mixture to reduce charging of samples. Homogeneity and uniformity of the sample are extremely important, particularly when using ion beam analysis techniques where the spot size of the beam is smaller than the sample to be irradiated.

5. Ion beam methods and synchrotron-based techniques in hair elemental profiling

Hair analysis has to date mostly exhausted conventional techniques that require digestion and hence destruction of a relatively large amount of hair. Such bulk analyses destroy information on the localized concentrations of minerals in a tissue. Ion beam and synchrotron radiation techniques have proven to be a solution to this dilemma. Numerous publications now exist that demonstrate the versatility of ion beam and synchrotron radiation analysis to map the quantitative distribution of minerals in hair cross sections or show the longitudinal distribution of elements in hair tissues [56, 57].

The unifying characteristics of these techniques include their sensitivity, selectivity, quantitative multielemental character, and speed of analysis [58]. Furthermore, depth profiling without physical sectioning is often possible with ion beam techniques, while the ion microprobe may allow μm or lower spatial resolution for diverse applications. Analysis of intact samples without any chemical processing or dissolution is also possible, whereas damage to analyzed samples may be minimal, allowing for downstream analyses with other techniques.

Ion beam and synchrotron radiation methods are based on the interaction of nuclear particles with atomic nuclei. Chemical bonding information is, however, not available as the electronic shell of the atom does not contribute to the physical process. Chemical speciation information is also not available with certain ion beam techniques, although results are not affected by the chemical form of the element. Furthermore, only surface near regions may be analyzed in most cases because of the short range of ions in matter. Perhaps the most significant downside of these techniques is the expensive equipment required and the fact that access to nuclear analytical facilities and expertise may not be accessible in some developing regions of the world.

Important considerations when performing nuclear analysis include the type of standard to use, the thickness of the sample, as well as the effect of irradiation on the sample [59]. The use of standards may vary depending on the type and aim of the study, while it is imperative to analyze relatively thin material (few nm to 10 μm) due to the energy loss of the charged particles. For irradiation effect, there are some detailed studies on whether analysis with ion beams lead to sample damage. To illustrate, in secondary ion mass spectrometry analysis at MeV energies (MeV-SIMS) and at low beam fluences—meaning the number of primary ions hitting the target area unit—the analysis should be nondestructive [60]. Over time, the yield of secondary molecules should also exponentially decrease as a function of beam fluence. The slope of this exponential fall is determined by the damage cross section, which explains the damage induced on the specimen surface by one primary ion. Experiments at the Jožef Stefan Institute in Slovenia utilizing a 5.8 MeV $^{35}Cl^{6+}$ primary ion beam reported damage cross section values of approx. 2 nm^2 for the amino acids arginine and leucine. A fluence of 10^{12} ions/cm^2 is the commonly accepted static limit of MeV-SIMS corresponding to 3 hours of measurement that agrees to approximately 8% of the target surface undergoing chemical alteration due to ion-induced damage. Since hair sample measurements is normally 1 hour, the hair chemical environment and morphology should remain unchanged. Scanning electron micrograph images further confirmed that the morphology of the analyzed samples remained unaltered after 1 hour of measurement with MeV-SIMS [53]. With particle-induced X-ray emission, however, a clear change in the sample's physical appearance can be noted when measuring for a longer time frame.

Besides sample damage, volatile analytes may also be lost during sample irradiation as a function of their thermal and radiation stability [61]. Adjusting the beam intensity and irradiation time may assist with alleviating this problem. Alternatively, another ion beam technique may be chosen to avoid this problem. For example, in-air proton-induced X-ray emission lends the advantage of performing measurements in air, resulting in negligible charging of the insulating targets, reduced radiation damage to the specimen due to the cooling effect of the air, and ultimately reduced loss of volatile elements [62]. Representative spectra obtained from a single hair fiber with in-air proton-induced X-ray emission is represented in **Figure 4**.

Another concern includes charge buildup on samples. For hair samples, charge buildup has been observed during bombardment with protons [55]. Many factors may influence this phenomenon, but most importantly, the thickness of the hair. Charge buildup and episodic discharging from samples lead to high background levels in the spectra and should be avoided at all cost. This may be achieved by blending thick targets with graphite or, alternatively, by coating thin samples with a layer of carbon or other appropriate conducting materials.

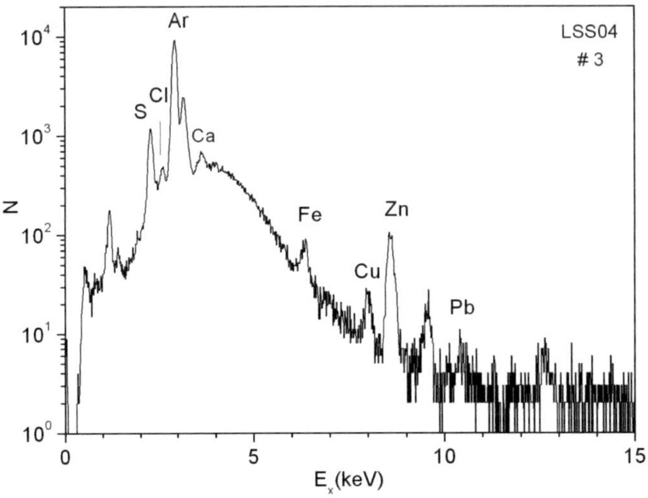

Figure 4. Representative in-air PIXE spectra from a single scalp hair, with the x-axis representing characteristic X-ray emission energy (KeV) and the y-axis representing X-ray emission intensity (counts) (unpublished data).

In the subsequent paragraphs, the focus will be placed on specific ion beam and synchrotron radiation analytical methods used in hair analyses. It should be borne in mind that techniques such as neutron activation analysis allow analysis of an entire sample, while specific ion beam techniques allow analysis of only the surface of a sample.

5.1. Neutron activation analysis

Neutron activation analysis has been described as a versatile method with a low detection limit for sample analysis of over 60 elements (major and trace elements) [63]. Samples do not require chemical preparation in which volatile elements may be lost. In addition, laborious sample homogenization procedures are not required for when a large amount of samples representative of a population need to be analyzed. The technique further requires only a small amount of sample weighing from 1 μm to hundreds of grams. Samples to be analyzed may also vary in form and shape. However, this need to be taken into consideration as sample geometry may affect results and also influence the choice of standard. A schematic overview of the technique is provided in **Figure 5**.

Instrumental neutron activation analysis is a popular analytical tool for the determination of elements in hair specifically for forensic applications [64]. Interestingly, instrumental neutron activation analysis has been applied in the forensic analysis of Napoleon's hair [65]. Napoleon died at the age of 51 on 5 May 1821, with his death officially ascribed to stomach cancer. Extensive investigations exploiting hair analysis, however, tried to dispute the officially declared cause of death as stomach cancer. Radiochemical neutron activation analysis showed that Napoleon's hair contained 10.38 ppm of arsenic. Longitudinal hair analysis to assess retrospective exposure to arsenic further showed that the arsenic was unevenly distributed along the hair length. Additional investigations with instrumental neutron activation analysis showed that Napoleon's hair also contained mercury, and abnormal levels of

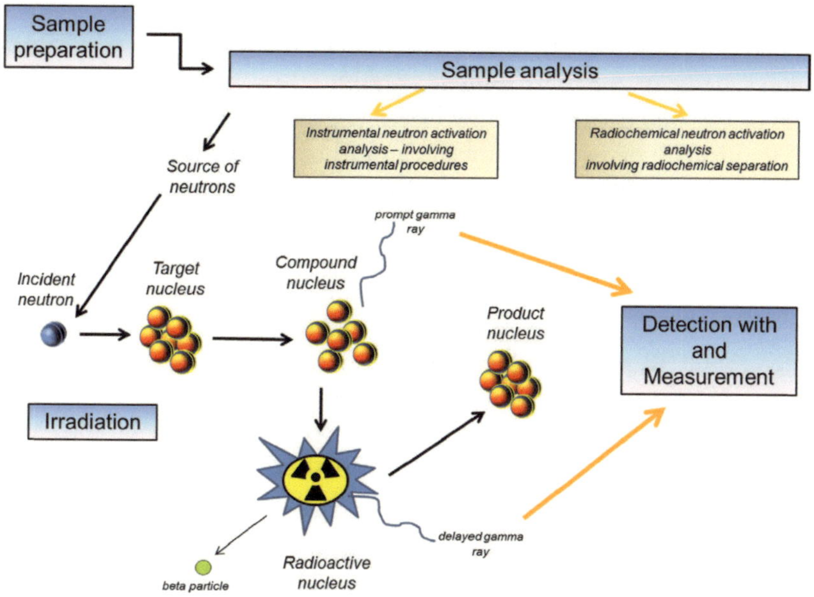

Figure 5. Schematic overview of the basic principle of neutron activation analysis.

chromium, antimony, and zinc that may have entered the body via food or medication. For example, Napoleon consumed a great deal of calomel that contains mercury to relieve constipation and thirst, while antimony was a component of tartar emetic, which was given to him shortly before his death as an agent to inhibit vomiting that would have expelled the poisons from his body.

Doubt, however, still exists whether the concentration of arsenic in Napoleon's hair was of endogenous or exogenous origin as exogenous contaminants may also be evenly distributed along the length of the hair. Hair may absorb arsenic from external contaminants that include wallpaper, coal smoke, water, cosmetics, and preservatives. The absorbed arsenic from either endogenous origin appearing in the medulla or exogenous origin appearing at the hair surface regions may remain in the hair structure even after washing. The conundrum of differentiating exogenous from endogenous exposure in hair can perhaps be more clearly understood through the use of ion beam imaging techniques such as particle-induced X-ray emission.

5.2. Ion beam imaging

Ion beam imaging has become a very popular tool for understanding the distribution of minerals in biological tissues. For hair elemental screening, particle-induced X-ray emission has emerged as one of the most popular and powerful ion beam techniques allowing quantitative elemental mapping in tissues at micrometer or lower resolution that well complements data obtained using other techniques such as synchrotron radiation X-ray fluorescence (XRF). The only disadvantage of the technique is that not all countries have robust facilities hosting the technique.

The technique falls under a broad family of X-ray emission techniques for the quantitative distribution mapping of low Z elements ($20 < Z < 35$ and $75 < Z < 85$) [66]. Particle-induced X-ray emission is based on the irradiation of a sample with a high-energy ion beam (typically 1–2 MeV of H or He) that excites the inner electron shells of an atom, inducing characteristic X-rays to be emitted by de-excitation of the atom. As negligible overlapping of characteristic X-rays for different elements occurs as the energy of an emitted X-ray is characteristic for a target element, multielemental detection becomes possible. X-rays are measured by an energy-dispersive (for example Si[Li]) detector placed at an angle of 135° relative to the beam direction to minimize the Bremsstrahlung background. When correcting for absorption and X-ray yields, quantitative data for elements may be obtained at detection limits of orders of magnitude around 10 ppm for thick samples. Since the physical processes involved in the generation of X-rays are well studied, no standards are required for producing quantitative elemental imaging with proton-induced X-ray emission. Elemental imaging in samples is obtained by scanning the ion beam microprobe over the surface of the specimen with a penetration depth of up to 100 μm.

Variants of particle-induced X-ray emission such as microproton-induced X-ray emission and in-air proton-induced X-ray emission offer supplementary benefits. For example, microproton-induced X-ray emission offers the added advantage of spatially resolved multielemental analysis of micrometer resolution [66], while in-air proton-induced X-ray emission lends the added advantage of performing measurements in air resulting in negligible charging of the insulating targets, reduced radiation damage and loss of volatile elements due to the cooling effect of the air, and simple handling and changing of samples [62]. In essence, ion beam techniques such as proton-induced X-ray emission provide nondestructive, fully quantitative, and multielemental mapping at micrometer-scale or lower spatial resolution and μg/g-level sensitivity of samples that require minimal processing [67]. Examples of applications of ion beam imaging using particle-induced X-ray emission for hair research will be discussed in Section 6. It is also recommended that the readers familiarize themselves with the numerous literatures describing the experimental setup, analysis, and data processing for the technique.

5.3. Time-of-flight MeV secondary ion mass spectrometry

Another technique that may well complement particle-induced X-ray emission is time-of-flight SIMS. Particle-induced X-ray emission may be used to quantify the elemental fingerprint of intact human scalp hair fibers, whereas time-of-flight SIMS may give an indication of both organic and inorganic compounds present in longitudinally sectioned hair. Time-of-flight SIMS is currently one of the most sensitive surface analysis techniques for chemical mapping, providing micrometer or lower resolution for depth profiling of the first one or two surface monolayers in an intact sample [47]. In addition, it allows parallel analyses and imaging of multiple elements, isotopes, and molecules in complex samples without exhaustive sample preparation such as labeling to provide information on the spatial localization of organic and inorganic compounds [68]. The main drawback of the technique is, however, the inability to quantify secondary ions due to the effect of the chemical composition of the matrix on the yield of secondary ions, also commonly referred to as the matrix effect [69].

http://dx.doi.org/10.5772/intechopen.72076

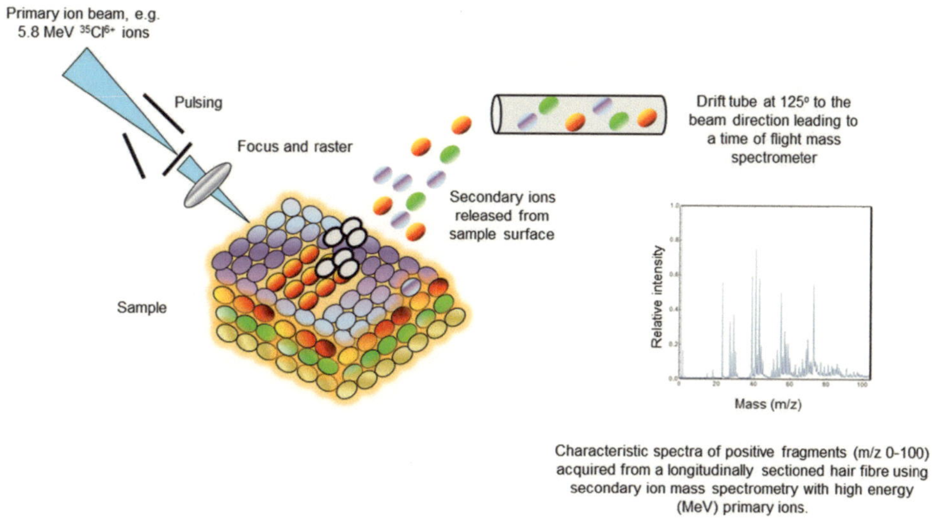

Figure 6. Schematic overview of the basic principle of time-of-flight secondary ion mass spectrometry with high-energy (MeV) primary ions.

In time-of-flight SIMS analysis, the sample is traditionally bombarded with ions in the energy range of 5–25 keV [68]. Recently, the technique has, however, also been tested with primary ions at MeV energies in a method termed MeV-SIMS [60, 70]. A schematic overview of the technique is provided in **Figure 6**. In MeV-SIMS, the interactions between primary ions and sample molecules are based on electronic energy loss, which creates a softer sputtering process and a 1000 times greater yield of larger and intact ionized molecules than conventional SIMS in the keV energy range. A pulse of specific primary ions at MeV energy—depending on the molecules of interest—is focused and rastered over the sample surface under ultra-high vacuum to generate position-dependent mass spectra [60, 70]. Various charged ions, also termed secondary ions, are released by the impact of the MeV primary ions with the sample surface. These secondary ions are of equal kinetic energy and enter a drift tube in which they are separated according to their mass-to-charge ratio. The masses of the secondary ions depend on the chemical composition of the sample scanned at each point, while the intensity of the secondary ion signal depends on the concentration of the particular compound in the area sampled as well as the ionization yield of the compound. The yield of secondary ions in turn depends on the incident angle, nature, and energy of the primary ion beam as well as the chemical properties of the target matrix.

The technique allows (1) parallel counting of secondary ions matched with preselected masses obtained from the analyzed points and (2) simultaneous building of chemical images [47]. The technique further allows for mass spectra to be extracted from an image region and vice versa to display the distribution of a secondary ion within the tissue. Interestingly, the technique is also able to analyze isotopic composition of light elements in a matrix such as hair [53]. The reader is, however, strongly recommended to first familiarize themselves with the experimental setup and a range of available detectors, the mechanism involved in secondary ion formation and the effect of the matrix composition on yield, as well as data processing and multivariate statistics deployed in SIMS before tackling this robust and challenging technique.

5.4. Synchrotron radiation

When electrons travel near the speed of light and are forced to change direction by a magnetic field, electromagnetic radiation is emitted that is also commonly described as synchrotron radiation [58, 71]. Today, synchrotron radiation has become one of the most powerful methodological tools for understanding the properties of matter in various scientific fields. The distinctive properties of synchrotron light include its ability to provide an energy tunable source of X-rays that can be highly linear, circular, or elliptically polarized; have a high brightness and collimation; have a wide energy spectrum with energies ranging from infrared light to hard X-rays; and provide nanosecond long light pulses that permit time-resolved studies.

Synchrotron facilities are characterized by a storage ring from which synchrotron radiation is emitted in a forward direction in a narrow cone that is at a tangent to the electron's orbit [71]. The width of this cone of electrons is affected by the speed of the electrons, which further affects the spectrum of the radiation, shifting it toward shorter wavelengths with increasing electron energy. The storage ring of so-called third-generation light sources is further equipped with magnetic insertion devices termed undulators and wigglers that are used to generate linear or circular polarized light by generating magnetic fields that drive electrons into an oscillating or spiral trajectory. These third-generation facilities may specialize in short-wavelength (hard X-rays), long-wavelength (soft X-rays), or vacuum-ultraviolet X-rays.

To focus the X-rays on the sample at micro- or nanometer resolution, different types of X-ray optics are used at various synchrotron facilities [72]. Some of these systems such as the Kirkpatrick-Baez mirror system use two glancing-angle, bowl-shaped mirrors with a curve designed to demarcate the beam profile at a focal point. Other systems include refractive lenses or Fresnel zone plates. These optical systems permit hard X-rays to be focused to 30–150 nm spot sizes that are injected to the sample. Ideally, the energy of the X-ray beam should be selected over a specified energy range. X-ray spectra are recovered subsequent to step- or raster scanning of the sample in the x and y planes at 45° to the beam path. More than one detector may be used to capture the spectra and deliver data with the beam x,y coordinates for each incident beam point. Conventional detectors for various applications that are common include Si(Li) detectors, intrinsic Ge detectors, or silicon drift detectors.

Probably the most important aspect of the analysis is optimizing the absorption of the X-rays in the samples to be analyzed [72]. The absorption of different types of X-rays produced in a synchrotron is controlled by the photoelectric effect. However, absorption may also be influenced by the chemical environment of the sample and element oxidation state. When an X-ray penetrates a sample and its energy is lower than the binding energy of the core electrons in the element of significance, its atoms do not absorb the X-rays. When the energy of the incoming X-rays and the binding energy of the core level electrons of an element are equal, X-ray photons are absorbed. Photoelectrons are produced since the incident X-ray electrons excite the core electrons. For the atom to return to a ground state, an electron needs to fill the empty position of the ejected photoelectron. This transition energy or wavelength is released as fluorescence or as an Auger electron for lighter elements. The number or intensity of X-ray photons is positively correlated with atomic abundance and hence individual element quantities. This is however mostly true for thin samples in which self-absorption within the sample is negligible. The net X-ray intensities of the elemental makeup are obtained via spectral deconvolution methods.

The absorption of X-rays into matter outlined above forms the backbone of a versatile synchrotron technique, X-ray absorption spectroscopy (XAS). The latter technique provides detailed information on elemental chemistry such as oxidation state and molecular geometry [72, 73]. Based on the energy region of interest, XAS can be further differentiated into X-ray absorption near-edge structure, extending from the pre-edge region to approximately 50 eV above the absorption edge and extended X-ray absorption fine structure (EXAFS), extending from about 50 to 1000 eV beyond the edge. Since electron transitions may occur from a partly bound and excited state in the pre-edge region, certain chemical features may become visible in this region. For EXAFS, the physical processes giving rise to the signal can be modeled by selected computer programs that allow accurate assessment of the identity of the surrounding atoms, bond distances, and coordination numbers.

Another technique based on the absorption of X-rays in matter described above is XRF [72, 73]. XRF permits quantitative elemental mapping. The physics of the photon interaction with matter is also well understood, simplifying quantification of data. Although XRF is similar to proton-induced X-ray emission and scanning electron microscopy with energy dispersive X-ray spectroscopy, synchrotron XRF is more sensitivity due to a high photon flux, weak scattering, and the availability of a tunable beam. Currently, XRF is one of the most sensitive imaging techniques offering spatially resolved down to 100 nm and quantitative topographical maps for a range of elements at submicron resolution. Depth resolution, however, depends on the elements of interest and sample nature, although high penetration depths of up to 1000 μm is possible that allows imaging of thicker samples or in-situ experiments.

Since this section only summarizes the background to synchrotron radiation and applicable techniques for analysis of hair samples, it is recommended that the readers familiarize themselves with the numerous literature sources available on specific synchrotron radiation techniques applicable to their samples and familiarize themselves with the beamlines and experimental setup at a chosen synchrotron facility. The latter may include the type of source device (bending magnet, wiggler magnet, or undulator), mirrors used to deflect or focus the X-rays, monochromators used for sorting incoming X-ray energies, glitch avoidance (double diffractions occurring in monochromator cryst detectors), detectors, availability of true imaging microscopes, electronics, and most importantly, data collection and statistics. Furthermore, it is important to familiarize oneself with an optimal sample preparation and calibration methodology for a specific synchrotron technique, elements accessible with the technique, effect of external parameters such as temperature on sample analysis, and nonscientifically, the culture of the facility and its researchers. Most importantly, beamtime application procedures also warrant careful attention.

6. Biomedical applications of ion beam and synchrotron analysis of hair

The amount of literature on hair chemical and specifically elemental analysis has grown substantially in recent years and with the specific focus on elemental analysis, it has been shown that almost any macro, trace, and xenobiotic elements can be quantitatively measured and mapped in hair tissues. The predominant amount of work with ion beam and synchrotron analyses has, however, focused on analyzing the distribution and content of elements in the hair of diseased patients versus healthy controls or those exposed to environmental pollutants. In the subsequent sections, examples of selected studies that employed ion beam and synchrotron radiation analysis to link elemental content with morbidity and toxicology will be discussed.

6.1. Dermatology

There is an alarming lack of fundamental studies describing the role of elements in the growth and physiology of hair or how elemental levels vary during the different growth stages of hair. Variation in elemental levels in hair at different stages of growth is also an important confounding variable in hair elemental research. The first researchers that tried to assess the quantitative elemental distribution in organ-cultured hair follicles during the anagen and catagen growth phases were from the Jožef Stefan Institute in Slovenia and ion microprobe facility ATOMKI in Debrecen [74]. Combined ion beam analysis exploiting proton-induced X-ray emission and scanning transmission ion microscopy were used to quantify elemental content in hair follicles during the catagen and anagen growth phases. The results showed that elemental concentrations were similar in selected parts of the hair follicle in both growth stages. However, the outer/inner root sheath keratinocyte layers of hair follicles in the catagen growth phase contained four times more calcium than the same regions in hair follicles in the anagen growth phase. These layers are known to express the receptor for capsaicin, TRPV1, which function as a calcium-permeable channel [75]. During the catagen growth phase of hair, an increase in the intracellular calcium concentration inhibits the proliferation of keratinocytes that triggers the induction of apoptosis [76]. In summary, this study showed the promise of ion beam techniques such as proton-induced X-ray emission and scanning transmission ion microscopy to improve our understanding of changes in elemental levels during hair growth.

6.2. Pediatrics

The deleterious effect of environmental pollutants such as lead on human health cannot be over emphasized [77]. The outcomes of lead exposure in children are particularly concerning due to its irreversible effects on one of the most sensitive organs to lead exposure, the developing brain [78]. In particular, children exposed to lead may have cognitive and behavioral effects that persist into adulthood [79]. To assess daily lead absorption during the prenatal phase, the longitudinal distribution of lead in fetal and parental hair was studied using synchrotron radiation micro-XRF [80]. The technique proved very successful in mapping lead along a longitudinal length that reflected a retrospective profile of lead exposure and absorption during the prenatal phase.

Hair elemental profiles may also find application as risk factors for disease. For example, intraindividual variations of 32 hair elemental levels as epidemiological risk factors for atopic dermatitis in infants were studied using particle-induced X-ray emission combined with rigorous regression statistics [57]. Hair samples were retrieved 1 month after birth and also at 10 months after birth with the onset of atopic dermatitis. The results showed that selenium and strontium could be used as explanatory variables in a regression model for atopic dermatitis. However, large intraindividual variations for selenium and strontium could affect the regression coefficients for strontium and selenium. Particle-induced X-ray emission was specifically employed in this study to understand intraindividual variations for selenium and strontium to correct for intraindividual variations in the previously described regression model. Such type of approach combining analytical techniques with statistical models is particularly important in promoting the application of hair elemental data in epidemiological research.

6.3. Psychiatry

Trace elements have also been shown to play an important role in psychiatric diseases. A study using proton-induced X-ray emission was conducted to assess the trace elements in the scalp hair of patients with alcohol-induced psychosis [81]. The results showed that iron and copper levels were higher in patients with alcohol-induced psychosis, whereas the concentrations of manganese and zinc were lower, compared to healthy controls.

The same approach was applied to identify and quantify elements in the scalp hair of patients with bipolar disorder based on gender [82]. For males, the concentration of copper was higher in bipolar patients compared to controls, whereas the concentrations of manganese, iron, zinc, and selenium were lower in bipolar patients of both genders compared to healthy controls. The same was observed in females, except that instead of manganese, nickel was higher in bipolar patients of both genders than controls. Furthermore, the Cu/Zn ratio was found to be higher in bipolar patients of both genders. This is not surprising as it is known that elemental dyshomeostasis triggers the formation of free radicals that may in turn affect neurotransmitter activity that plays an important role in psychiatric disorders [83]. However, treatment may also affect elemental levels in psychiatric patients, emphasizing the importance of assessing elemental levels before and after treatment [84].

6.4. Oncology

Breast cancer is a devastating cancer affecting the lives of many women and men around the world. Interestingly, hair samples have been used to assess the risk of breast cancer [85]. With XRF and X-ray diffraction techniques, it was shown that not only elemental levels but also hair structure differ between healthy controls and those with breast cancer [86]. More specifically, the data revealed that trace element levels were higher in healthy controls than in breast cancer patients and that wavelength of XRF presented with a 96% sensitivity compared to a 77% sensitivity for mammography, the gold standard for breast cancer screening.

Besides breast cancer, the effect of radiation therapy on hair trace elemental concentrations in cervical cancer patients has also been investigated with ion beam analysis; in this case, proton-induced X-ray emission [87]. Testing was done before and during radiation therapy. The concentrations of chlorine, potassium, calcium, titanium, chromium, manganese, iron, nickel, and zinc were lower and copper higher before irradiation. It was also shown that the concentration of specific elements varied during the course of radiation therapy, indicating a possible effect of this type of cancer therapy on elemental levels in the body.

6.5. Toxicology: environmental and occupational exposure to heavy metal pollution

Elemental analysis of hair samples also now finds important applications in the fields of toxicology. In this field, ion beam techniques such as particle-induced X-ray emission has been mostly used in the evaluation of toxic element pollution in various regions of the world [88]. For example, one study used particle-induced X-ray emission to assess the levels of toxic elements in the vicinity of a mining area in Mongolia, which included places of milling, grasslands, and villages [89]. Among the samples tested, human hair was also used. The average concentrations of titanium, arsenic, and strontium were found to be higher in hair samples from miners than that of control participants. Besides studies related to environmental pollution, ion beam analysis has also been applied in toxicology research. For example in toxicology studies, mercury has completely overpopulated the literature [14].

6.6. Veterinary science

Elemental analysis of hair samples has also branched into the field of veterinary sciences. Interestingly, ion beam techniques have also been exploited in animal hair analysis. For example, particle-induced X-ray emission was employed to assess the levels of elements in the main hair of horses in a study aimed at investigating the relationship between main hair elemental concentrations and the severity of second degree atrioventricular block in horses [90]. The results showed that there was a significant positive correlation between the zinc/copper ratio and calcium concentration in main hair and the drop in ventricular beats measured hourly in animals with a second-degree atrioventricular block. Receiver operating characteristic curve analysis suggested that the cut-off points for hair calcium concentration were set at 1536 μg/g and 26.0 for the zinc/copper ratio in detecting second-degree atrioventricular block in horses. It was concluded that the levels of calcium, zinc, and copper in the main hair of horses may be used as a diagnostic tool for testing susceptibility to atrioventricular block.

7. Summary

Today, more and more studies connect metabolic homeostasis and morbidity with the complicated interaction between essential and nonessential elements [91]. However, disease may also perturb elemental levels and alter the expression of other chemical compounds in the body [14]. A good example here includes the effect of arsenic exposure on the reduction of glucose tolerance [92]. Elemental dyshomeostasis in the body will also affect hair metabolic processes and hence elemental levels within the actively growing hair.

A fundamental understanding of what for example elemental concentrations in hair actually describe still requires more rigorous research and debate. More specifically, more studies investigating the incorporation of chemicals into hair, for example how elemental levels relate to blood levels and other biomarkers of disease, and most importantly, the contribution of contaminants to biogenic hair elemental content, are required. When performing hair analysis, careful consideration of the various confounding factors in hair elemental analysis such as medication use, smoking, and seasonal diet fluctuations is also warranted [14, 93]. In essence, these considerations are imperative before the intricate link can be cemented between hair elemental concentrations and its biological significance related to morbidity or other factors.

Studies utilizing ion beam and synchrotron radiation techniques have, however, improved our understanding of the distribution of elements in hair tissue and elemental dyshomeostasis in disease [50, 80, 82]. Hair elemental analysis finds perhaps the most important application in toxicology research relating to acute versus chronic exposure to toxic metals and other xenobiotic compounds [80]. A fascinating example includes the case of the famous race horse Phar Lap in which synchrotron radiation XRF was used to confirm the cause of death [94]. Arsenic mapping in hair using synchrotron XRF and XAS not only allowed longitudinal distribution mapping of arsenic in the hair of Phar Lap, but also the retrospective changes in the metabolic incorporation of arsenic into the hair shaft. Another example includes the use of ion beam or synchrotron radiation analysis in epidemiological and etiological studies assessing exposure to toxic metals in vulnerable populations [14]. A recent exploratory study also pointed to the use of MeV-SIMS for the detection and mapping of the micronutrient lithium and its isotopes in longitudinally sectioned hair [53]. Lithium is an important psychopharmaceutical in the psychiatric community, and since it is quite challenging detecting the low Z element lithium that is present at trace levels in biological tissues, such analysis may find important application in monitoring medication adherence among psychiatric patients using hair. In summary, hair analysis remains a cost-effective baseline tool for not only more comprehensive studies linking morbidity with the chemicals and more specifically elements found in hair and vice versa, but also various other important applications in diverse fields. With the continuous development and optimization of ion beam and synchrotron radiation techniques, the future of hair analysis remains bright.

Author details

Karen J. Cloete

Address all correspondence to: kaboutercloete@gmail.com

iThemba Laboratory for Accelerator Based Sciences-National Research Foundation, Somerset West, South Africa

References

[1] Soetan KO, Olaiya CO, Oyewole OE. The importance of mineral elements for humans, domestic animals and plants: A review. African Journal of Food Science. 2010;**4**:200-222

[2] Mann J, Truswell A, editors. Essentials of Human Nutrition. 5th ed. Oxford: Oxford University Press; 2017. 720 p

[3] Prashanth L, Kattapagari KK, Chitturi RT, Baddam VR, Prasad LK. A review on role of essential trace elements in health and disease. Journal Dr. NTR University of Health Sciences. 2015;**4**:75-85. DOI: 10.4103/2277-8632.158577

[4] Zhang X-Y, Zheng L-N, Wang H-L, Shi J-W, Feng W-Y, Li L, Wang M. Elemental bio-imaging of biological samples by laser ablation-inductively coupled plasma-mass spectrometry. Chinese Journal of Analytical Chemistry. 2016;**44**:1646-1651. DOI: 10.1016/S1872-2040(16)60969-6

[5] Hwang C, Hong K. Other micronutrient deficiencies in inflammatory bowel disease: From A to Zinc. In: Ananthakrishnan AN, editor. Nutritional Management of Inflammatory Bowel Diseases. Switzerland: Springer; 2016. p. 65-101. DOI: 10.1007/978-3-319-26890-3

[6] Dinis-Oliveira RJ, Carvalho F, Duarte JA, Remião F, Marques A, Santos A, Magalhães T. Collection of biological samples in forensic toxicology. Toxicology Mechanisms and Methods. 2010;**20**:363-414. DOI: 10.3109/15376516.2010.497976

[7] Takai N, Tanaka Y, Inazawa K, Saji H. Quantitative analysis of pharmaceutical drug distribution in multiple organs by imaging mass spectrometry. Rapid Communications in Mass Spectrometry. 2012;**26**:1549-1556. DOI: 10.1002/rcm.6256

[8] Norris JL, Caprioli RM. Analysis of tissue specimens by matrix-assisted laser desorption/ionization imaging mass spectrometry in biological and clinical research. Chemical Reviews. 2013;**113**:2309-2342. DOI: 10.1021/cr3004295

[9] Mercolini L, Protti M. Biosampling strategies for emerging drugs of abuse: Towards the future of toxicological and forensic analysis. Journal of Pharmaceutical and Biomedical Analysis. 2016;**130**:202-219. DOI: 10.1016/j.jpba.2016.06.046

[10] El BG, Flieger J, Huber M, Kocjan R. Examination of the elemental composition of hair in cholelithiasis, kidney stone, hypertension and diabetes by scanning electron microscopy and energy dispersive spectrometry SEM/EDS. Journal of Analytical & Bioanalytical Techniques. 2014;**5**:207. DOI: 10.4172/2155-9872.1000207

[11] Schöpfer J, Schrauze GN. Lithium and other elements in scalp hair of residents of Tokyo prefecture as investigational predictors of suicide risk. Biological Trace Element Research. 2011;**144**:418-425. DOI: 10.1007/s12011-011-9114-x

[12] Zaitseva IP, Skalny AA, Tinkov AA, Berezkina ES, Grabeklis AR, Skalny AV. The influence of physical activity on hair toxic and essential trace element content in male and female students. Biological Trace Element Research. 2015;**163**:58-66. DOI: 10.1007/s12011-014-0172-8

[13] Abbruzzini TF, Silva CA, Andrade DA, Carneiro WJO. Influence of digestion methods on the recovery of iron, zinc, nickel, chromium, cadmium and lead contents in 11 organic residues. Revista Brasileira de Ciência do Solo. 2014;**38**:166-176. DOI: 10.1590/S0100-06832014000100016

[14] Kempson IM, Lombi E. Hair analysis as a biomonitor for toxicology, disease and health status. Chemical Society Reviews. 2011;**40**:3915-3940. DOI: 10.1039/c1cs15021a

[15] Jaber JA, Holt D, Johnston A. Method development for the detection of basic/weak basic drugs in hair by LCMSMS: Comparison between methanolic and alkaline extraction on real samples. Pharmacology & Pharmacy. 2012;**3**:263-274. DOI: 10.4236/pp.2012.33035

[16] Namera A, Kawamura M, Nakamoto A, Saito T, Nagao M. Comprehensive review of the detection methods for synthetic cannabinoids and cathinones. Forensic Toxicology. 2015;**33**:175-194. DOI: 10.1007/s11419-015-0270-0

[17] Pillay AE. A review of accelerator-based techniques in analytical studies. Journal of Radioanalytical and Nuclear Chemistry. 2000;**243**: 191-197. DOI: 10.1023/A:1006748104841

[18] Nastasi M, Mayer JW, Wang Y, editors. Ion Beam Analysis, Fundamentals and Applications. Boca Raton: CRC press; 2014. 472 p

[19] Jeynes C, Webb RP, Lohstroh A. Ion beam analysis: A century of exploiting the electronic and nuclear structure of the atom for materials characterisation. Reviews of Accelerator Science and Technology. 2011;**4**:41-82. DOI: 10.1142/S1793626811000483

[20] Dong Z, Jim RC, Hatley EL, Backus ASN, Shine JP, Spengler JD, Schaider LA. A longitudinal study of mercury exposure associated with consumption of freshwater fish from a reservoir in rural South Central USA. Environmental Research. 2015;**136**:155-162. DOI: 10.1016/j.envres.2014.09.029

[21] Favretto D, Vogliardi S, Tucci M, Simoncello I, El Mazloum R, Snenghi R. Occupational exposure to ketamine detected by hair analysis: Retrospective and prospective toxicological study. Forensic Science International. 2016;**265**:193-199. DOI: 10.1016/j.forsciint.2016.03.010

[22] Preedy V, editor. Handbook of Hair in Health and Disease. The Netherlands: Wageningen Academic Publishers; 2012. 476 p. DOI: 10.3920/978-90-8686-728-8

[23] Katz SA, Chatt A. The Use of Hair as a Biopsy Tissue for Trace Elements in the Human Body, Application of Hair as an Indicator for Trace Element Exposure in Man, A Review. NAHRES-22. Vienna: International Atomic Energy Agency; 1994. 11 p

[24] Rosen EP, Thompson CG, Bokhart MT, Prince HM, Sykes C, Muddiman DC, Kashuba AD. Analysis of antiretrovirals in single hair strands for evaluation of drug adherence with infrared-matrix assisted laser desorption electrospray ionization mass spectrometry imaging. Analytical Chemistry. 2016;**88**:1336-1344. DOI: 10.1021/acs.analchem.5b03794

[25] Buononato EV, De Luca D, Galeandro IC, Congedo ML, Cavone D, Intranuovo G, Guastadisegno CM, Corrado V, Ferri GM. Assessment of environmental and occupational exposure to heavy metals in Taranto and other provinces of Southern Italy by means of scalp hair analysis. Environmental Monitoring and Assessment. 2016;**188**:337. DOI: 10.1007/s10661-016-5311-6

[26] Kintz P. Forensic Science Series: Analytical and Practical Aspects of Drug Testing in Hair. Florida: Boca Raton; 2007. 382 p

[27] Pragst F, Balikova MA. State of the art in hair analysis for detection of drug and alcohol abuse. Clinica Chimica Acta. 2006;**370**:17-49. DOI: 10.1016/j.cca.2006.02.019

[28] Kegler R, Büttner A, Nowotnik J, Rentsch D. Postmortem investigation of 88-cm-long dreadlocks for drugs of abuse: An unusual case report in the northeast of Germany. Forensic Toxicology. 2016;**34**:419-424. DOI: 10.1007/s11419-016-0318-9

[29] Rasmussen KL, Kučera J, Skytte L, Kameník J, Havránek V, Smolík J, Velemínský P, Lynnerup N, Bruzek J, Velley J. Was he murdered or was he not? – Part I: Analysis of mercury in the remains of Tycho Brahe. Archaeometry. 2013;**55**:1187-1195. DOI: 10.1111/j.1475-4754.2012.00729.x

[30] Robbins CR, editor. Chemical and Physical Behavior of Human Hair. 4th ed.. New York: Springer; 2013. 483 p. DOI: 10.1007/b97447

[31] Gerace E, Veronesi A, Martra G, Salomone A, Vincent M. Study of cocaine incorporation in hair damaged by cosmetic treatments. Forensic Chemistry. 2017;**3**:69-73

[32] Dias MFRG. Hair cosmetics: An overview. International Journal of Trichology. 2015;**7**:2-15. DOI: 10.4103/0974-7753.153450

[33] Galliano A, Ye C, Su F, Wang C, Wang Y, Liu C, Wagle A, Guerin M, Flament F, Steel A. Particulate matter (PM's) adheres to human hair exposed to severe aerial pollution: Consequences for certain hair surface properties. International Journal of Cosmetic Science. DOI: 10.1111/ics.12416

[34] Brewer PA, Bird G, Macklin MG. Isotopic provenancing of Pb in Mitrovica, northern Kosovo: Source identification of chronic Pb enrichment in soils, house dust and scalp hair. Applied Geochemistry. 2016;**64**:164-175. DOI: 10.1016/j.apgeochem.2015.08.003

[35] Robbins CR. Chemical Composition of Different Hair Types. 5th ed. New York: Springer; 2012. DOI: 10.1007/978-3-642-25611-0_2

[36] Kanti V, Nuwayhid R, Lindner J, Hillmann K, Stroux A, Bangemann N, Kleine-Tebbe A, Blume-Peytavi U, Garcia Bartels N. Analysis of quantitative changes in hair growth during treatment with chemotherapy or tamoxifen in patients with breast cancer: A cohort study. The British Journal of Dermatology. 2014;**170**:643-650. DOI: 10.1111/bjd.12716

[37] Grosvenor AJ, Marsh J, Thomas A, Vernon JA, Harland DP, Clerens S, Dyer JM. Oxidative modification in human hair: The effect of the levels of Cu (II) ions, UV exposure and hair pigmentation. Photochemistry and Photobiology. 2016;**92**:144-149. DOI: 10.1111/php.12537

[38] Baker JA, Ayad FK, Maitham SA. Influence of various parameters on the levels of arsenic in washed scalp hair from Karbala, Iraq by using ICP-OES technique. Karbala International Journal of Modern Science. 2016;**2**:104-112. DOI: 10.1016/j.kijoms.2016.02.004

[39] Błażewicz A, Liao KY, Liao HH, Niziński P, Komsta Ł, Momčilović B, Jabłońska-Czapla M, Michalski R, Prystupa A, Sak JJ, Kocjan R. Alterations of hair and nail content of selected trace elements in nonoccupationally exposed patients with chronic depression from different geographical regions. BioMed Research International. DOI: 10.1155/2017/3178784

[40] Skalny AV, Simashkova NV, Skalnaya AA, Klyushnik TP, Bjørklund G, Skalnaya MG, Tinkov AA. Assessment of gender and age effects on serum and hair trace element levels in children with autism spectrum disorder. Metabolic Brain Disease. DOI: 10.1007/s11011-017-0056-7

[41] Velasco MVR, de Sá Dias TC, de Freitas AZ, Júnior NDV, de Oliveira Pinto CAS, Kaneko TM, Baby AR. Hair fiber characteristics and methods to evaluate hair physical and mechanical properties. Brazilian Journal of Pharmaceutical Sciences 2009;**45**:153-162. DOI: 10.1590/S1984-82502009000100019

[42] Sukumar A. Factors influencing levels of trace elements in human hair. Reviews of Environmental Contamination and Toxicology. 2002;**175**:47-78

[43] Chojnacka K, Michalak I, Zielinska A, Gorecka H, Gorecki H. Inter-relationship between elements in human hair: The effect of gender. Ecotoxicology and Environmental Safety. 2010;**73**:2022-2028. DOI: 10.1016/j.ecoenv.2010.09.004

[44] Lemasters JJ, Ramshesh VK, Lovelace GL, Lim J, Wright GD, Harland D, Dawson Jr TL. Compartmentation of mitochondrial and oxidative metabolism in growing hair follicles: A ring of fire. The Journal of Investigative Dermatology. 2017;**137**:1434-1444. DOI: 10.1016/j.jid.2017.02.983

[45] Han E, Yang H, Seol I, Park Y, Lee B, Song JM. Segmental hair analysis and estimation of methamphetamine use pattern. International Journal of Legal Medicine. 2013;**127**:405-411. DOI: 10.1007/s00414-012-0766-7

[46] Poetzsch M, Baumgartner MR, Steuer AE, Kraemer T. Segmental hair analysis for differentiation of tilidine intake from external contamination using LC-ESI-MS/MS and MALDI-MS/MS imaging. Drug Testing and Analysis. 2015;**7**:143-149. DOI: 10.1002/dta.1674

[47] Vickerman JC, Briggs D. ToF-SIMS: Materials Analysis by Mass Spectrometry. Surface Spectra. 2nd ed. Chichester: Manchester and IM Publications, Chichester; 2013. 732 p

[48] Schaffer M, Hill V, Cairns T. Hair analysis for cocaine: The requirement for effective wash procedures and effects of drug concentration and hair porosity in contamination and decontamination. Journal of Analytical Toxicology. 2005;**29**:319-326

[49] Kempson IM, Skinner WM. A comparison of washing methods for hair mineral analysis: Internal versus external effects. Biological Trace Element Research. 2012;**150**:10-14. DOI: 10.1007/s12011-012-9456-z

[50] Pineda-Vargas CA, Eisa MEM. Analysis of human hair cross sections from two different population groups by nuclear microscopy. Nuclear Instruments and Methods in Physics Research B. 2010;**268**:2164-2167. DOI: 10.1016/j.nimb.2010.02.041

[51] Kempson IM, Skinner WM. ToF-SIMS analysis of elemental distributions in human hair. Science of the Total Environment. 2005;**338**:213-227. DOI: 10.1039/c1cs15021a

[52] Kempson IM, Skinner WM, Kirkbride PK. A method for the longitudinal sectioning of single hair samples. Journal of Forensic Sciences. 2002;**47**:889-892. DOI: 10.1520/JFS15452J

[53] Cloete KJ, Jenčič B, Šmit Ž, Kelemen M, Mkentane K, Pelicon P. Detection of lithium in scalp hair by time-of flight secondary ion mass spectrometry with high energy (MeV) primary ions. Analytical Methods. DOI: 10.1039/C7AY01616F

[54] Van Baar HM, Schalkwijk J, Perret CM. Cryosectioning of hair follicles. An improved method using liquid nitrogen conduction freezing. Acta Dermato-Venereologica. 1990; **70**:335-338

[55] Valkovic V.Sample Preparation Techniques in Trace Element Analysis by X-ray Emission Spectroscopy. IAEA-TECDOC-300. Vienna: International Atomic Energy Agency; 1983. 188 p

[56] Israelsson A, Eriksson M, Pettersson HBL. On the distribution of uranium in hair: Non-destructive analysis using synchrotron radiation induced X-ray fluorescence micro-probe techniques. Spectrochimica Acta B. 2015;**108**:28-34. DOI: 10.1016/j.sab.2015.04.001

[57] Yamada T, Saunders T, Nakamura T, Sera K, Nose Y. On intra-individual variations in hair minerals in relation to epidemiological risk assessment of atopic dermatitis. In: Kitsos C, Oliveira T, Rigas A, Gulati S, editors. Theory and Practice of Risk Assessment. Vol 136. Springer Proceedings in Mathematics and Statistics. Cham.: Springer; 2015. DOI: 10.1007/978-3-319-18029-8_9

[58] da Cunha MML, Trepout S, Messaoudi C, Wub T-D, Ortega R, Guerquin-Kern J-L, Marco S. Overview of chemical imaging methods to address biological questions. Micron 2016;**84**:23-36. DOI: 10.1016/j.micron.2016.02.005

[59] Verma HR. Atomic and Nuclear Analytical Methods XRF, Mössbauer, XPS, NAA and Ion Beam Spectroscopic Techniques. Berlin: Springer-Verlag; 2007. 376 p. DOI: 10.1007/978-3-540-30279-7

[60] Jenčič B, Jeromel L, Potočik LA, Vogel-Mikuš K, Kovačec E, Regvar M, Siketić Z, Vavpetič P, Rupnik Z, Bučar K, Kelemen M, Kovač J, Pelicon P. Molecular imaging of cannabis leaf tissue with MeV-SIMS method. Nuclear Instruments and Methods in Physics Research B. 2016;**371**:205-210. DOI: 10.1016/j.nimb.2015.10.047

[61] Lucarelli F, Calzolai G, Chiari M, Nava S, Carraresi L. Study of atmospheric aerosols by IBA techniques: The LABEC experience. Nuclear Instruments and Methods in Physics Research B. (in press). DOI: 10.1016/j.nimb.2017.07.034

[62] Giuntini L. A review of external microbeams for ion beam analyses. Analytical and Bioanalytical Chemistry. 2011;**401**:785-793. DOI: 10.1007/s00216-011-4889-3

[63] Witkowska E, Szczepaniak K, Biziuk M. Some applications of neutron activation analysis. Journal of Radioanalytical and Nuclear Chemistry. 2005;**265**:141-150. DOI: https://doi.org/10.1007/s10967-005-0799-1

[64] Vasconcellos MBA, Bode P, Paletti G, Catharino MGM, Ammerlaan AK, Saiki M, Fávaro DIT, Byrne AR, Baruzzi R, Rodrigues DA. Determination of mercury and selenium in hair samples of Brazilian Indian populations living in the Amazonic Region by NAA. Journal of Radioanalytical and Nuclear Chemistry. 2000;**244**:81-85. DOI: 10.1023/A:1006775006509

[65] Lin X, Alber D, Henkelmann R. Elemental contents in Napoleon's hair cut before and after his death: Did Napoleon die of arsenic poisoning? Analytical and Bioanalytical Chemistry. 2004;**379**:218-220. DOI: https://doi.org/10.1007/s00216-004-2536-y

[66] Johansson SAE, Campbell JL, Malmqvist KG. Particle Induced X-Ray Spectrometry (PIXE). New York: Wiley; 1995

[67] Campbell JL. Particle-Induced X-Ray Emission. Characterization of Materials. New York: John Wiley and Sons; 2002

[68] Liu J, Ouyang Z. Mass spectrometry imaging for biomedical applications. Analytical and Bioanalytical Chemistry. 2013;**405**:5645-5653. DOI: 10.1007/s00216-013-6916-z

[69] Alnajeebi AM, Vickerman JC, Lockyer NP. Matrix effects in biological SIMS using cluster ion beams of different chemical composition. Biointerphases. 2016;**11**:02A317. DOI: 10.1116/1.4941009

[70] Jeromel L, Siketić Z, Potočnik NO, Vavpetič P, Rupnik Z, Bučar K, Pelicon P. Development of mass spectrometry by high energy focused heavy ion beam: MeV SIMS with 8 MeV Cl^{7+} beam. Nuclear Instruments and Methods in Physics Research B. 2014;**332**:22-27. DOI: 10.1016/j.nimb.2014.02.022

[71] Willmott P. An Introduction to Synchrotron Radiation: Techniques and Applications. New York. John Wiley and Sons; 2011. 368 p. DOI: 10.1002/9781119970958

[72] Collingwood JF, Adams F. Chemical imaging analysis of the brain with X-ray methods. Spectrochimica Acta Part B. 2017;**130**:101-118. DOI: 10.1016/j.sab.2017.02.013

[73] West M, Ellis AT, Potts PJ, Streli C, Vanhoofe C, Wobrauschekc P. Atomic spectrometry update – A review of advances in X-ray fluorescence spectrometry and its applications. Journal of Analytical Atomic Spectrometry. 2016;**31**:1706-1755. DOI: 10.1039/c6ja90034h

[74] Kertesz Z, Szikszai Z, Pelicon P, Simcic J, Telek A, Biro T. Ion beam microanalysis of human hair follicles. Nuclear Instruments and Methods in Physics Research B. 2007;**260**: 218-221. DOI: 10.1016/j.nimb.2007.02.025

[75] Bodó E, Bíró T, Telek A, Czifra G, Griger Z, Tóth BI, Mescalchin A, Ito T, Bettermann A, Kovács L, Paus R. A hot new twist to hair biology involvement of vanilloid receptor-1 (VR1/TRPV1) signaling in human hair growth control. The American Journal of Pathology. 2005;**166**:985-998. DOI: 10.1016/S0002-9440(10)62320-6

[76] Bikle DD. Vitamin D and the skin: Physiology and pathophysiology. Reviews in Endocrine & Metabolic Disorders. 2012;**13**:3-19. DOI: 10.1007/s11154-011-9194-0

[77] Mason LH, Harp JP, Han DY. Pb neurotoxicity: Neuropsychological effects of lead toxicity. BioMed Research International. 2014. DOI: 10.1155/2014/840547

[78] Ordemann JM, Austin RN. Lead neurotoxicity: Exploring the potential impact of lead substitution in zinc-finger proteins on mental health. Metallomics. 2016;**8**:579-588. DOI: 10.1039/c5mt00300h

[79] Counter SA, Buchanan LH, Ortega F. Neurocognitive status of Andean children with chronic environmental lead exposure. Journal of Environmental and Occupational Science. 2015;**4**:179-184. DOI: 10.5455/jeos.20151029110613

[80] Tong Y, Sun H, Luo Q, Feng J, Liu X, Liang F, Yan F, Yang K, Yu X, Li Y, Chen J. Study of lead level during pregnancy by application of synchrotron radiation micro XRF. Biological Trace Element Research. 2011;**142**:380-387. DOI: 10.1007/s12011-010-8805-z

[81] Pradeep AS, Nagaraju GJ, Sarita P. Trace elements in the scalp hair of patients with alcohol induced psychosis. Journal of Radioanalytical and Nuclear Chemistry. 2012;**294**:271-276. DOI: 10.1007/s10967-011-1591-z

[82] Pradeep AS, Naga Raju GJ, Sattar SA, Sarita P, Prasada Rao AD, Ray DK, Reddy BS, Reddy SB. Trace elemental distribution in the scalp hair of bipolars using PIXE technique. Medical Hypotheses. 2014;**82**:470-477. DOI: 10.1016/j.mehy.2014.01.028

[83] Gadoth N, Göbel HH. Oxidative Stress and Free Radical Damage in Neurology. New York: Springer; 2011. 323 p. DOI: 10.1007/978-1-60327-514-9

[84] Kazi TG, Baloch S, Afridi HI, Talpur FN, Sahito OM. Variation of lithium contents in scalp hair samples of different male psychiatric patients before and after treatment with its pharmaceutical supplements. Pharmaceutica Analytica Acta. 2016;**7**:520-526. DOI: 10.4172/2153-2435.1000520

[85] Adkins-Jackson SL, PB CP, Mitchell E, Montgomery S. A review of hair product use on breast cancer risk in African American women. Cancer Medicine. 2016;**5**:597-604. DOI: 10.1002/cam4.613

[86] Maziar A, Shahbazi-Gahrouei D, Tavakoli MB, Changizi V, Non-invasive XRF. Analysis of human hair for health state determination of breast tissue. International Journal of Cancer Management. 2015;**8**:e3983. DOI: 10.17795/ijcp-3983

[87] Charles MJ, Reddy SB, Raju GJN, Ravi Kumar M, Seetharamireddy B, Mallikharjuna Rao B, Reddy TS, Murty GAVR, Ramakrishna Y, Vijayan V, Ramani A. Effect of radiation therapy on trace elemental concentrations of hair samples of cervical cancer patients — PIXE technique. X-Ray Spectrometry. 2004;**33**:410-413. DOI: 10.1002/xrs.739

[88] Ejaz S, Camer GA, Anwar K, Ashraf M. Monitoring impacts of air pollution: PIXE analysis and histopathological modalities in evaluating relative risks of elemental contamination. Ecotoxicology. 2014;**23**:357-369. DOI: 10.1007/s10646-014-1193-y

[89] Oyuntsetseg B, Kawasaki K, Watanabe M, Ochirbat B. Evaluation of the pollution by toxic elements around the small-scale mining area, Boroo, Mongolia. ISRN Analytical Chemistry. 2012. DOI: 10.5402/2012/153081

[90] Suzuki K, Yamaya Y, Asano K, Chiba M, Sera K, Matsumoto T, Sakai T, Asano R.Relationship between hair elements and severity of atrioventricular block in horses. Biological Trace Element Research. 2007;**115**:255-264. DOI: 10.1007/BF02686000

[91] Chellan P, Sadler PJ. The elements of life and medicines. Philosophical Transactions. Series A, Mathematical, Physical, and Engineering Sciences. 2015;**373**(2037). DOI: 10.1098/rsta.2014.0182

[92] Farzan SF, Gossai A, Chasan-Taber YCL, Baker E, Karagas M. Maternal arsenic exposure and gestational diabetes and glucose intolerance in the New Hampshire birth cohort study. Environmental Health. 2016;**15**:106. DOI: 10.1186/s12940-016-0194-0

[93] Afridi HI, Talpur FN, Kazi TG, Brabazon D. Assessment of toxic elements in the samples of different cigarettes and their effect on the essential elemental status in the biological samples of Irish hypertensive consumers. Journal of Human Hypertension. 2015;**29**:309-315. DOI: 10.1038/jhh.2014.87

[94] Kempson IM, Dermot HA. Titelbild: Determination of arsenic poisoning and metabolism in hair by synchrotron radiation: The case of Phar Lap (Angew. Chem. 25/2010). Angewandte Chemie. 2010;**122**:4239. DOI: 10.1002/ange.201002029